Fallacy, Fraud, & Failure

At The Hands Of

Science, Medical, Dental,

& Other Authority

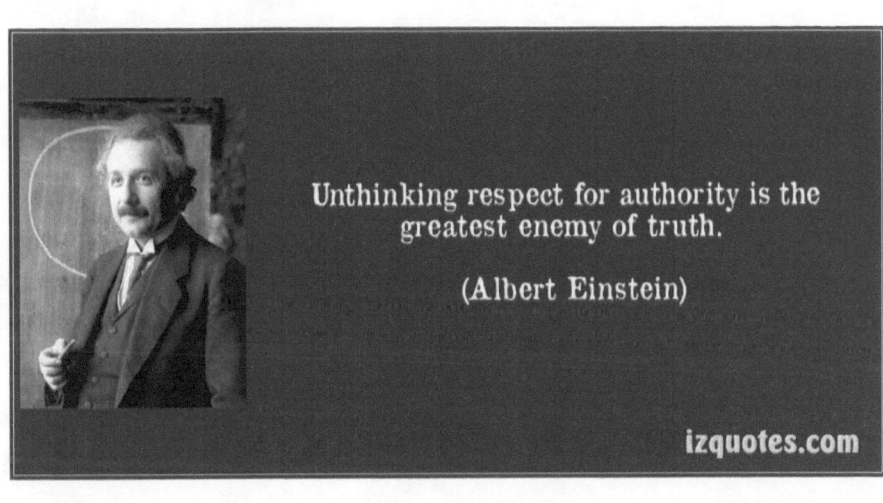

Unthinking respect for authority is the greatest enemy of truth.

(Albert Einstein)

izquotes.com

"I've been a two-pack-a-day man for fifteen years and I've found much milder Chesterfield is **best** for me."

Perry Como

NOW...10 Months Scientific Evidence For Chesterfield

A MEDICAL SPECIALIST is making regular bi-monthly examinations of a group of people from various walks of life. 45 percent of this group have smoked Chesterfield for an average of over ten years.

After ten months, the medical specialist reports that he observed...

no adverse effects on the nose, throat and sinuses of the group from smoking Chesterfield.

First and Only Premium Quality Cigarette in Both Regular and King-Size

MUCH MILDER
CHESTERFIELD
IS BEST FOR YOU

Copyright 1953, LIGGETT & MYERS TOBACCO CO.

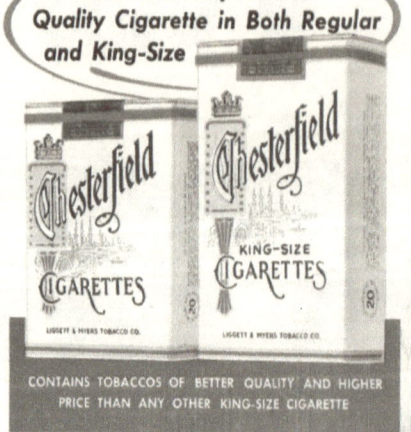

CONTAINS TOBACCOS OF BETTER QUALITY AND HIGHER PRICE THAN ANY OTHER KING-SIZE CIGARETTE

TOUTED AS 'SCIENTIIC EVIDENCE, READ ON.

AN EXAMPLE, TO ILLUSTRATE THE NEFARIOUS NATURE OF THE EXPERTS & AUTHORITY INVOLVED IN THIS ACTUAL CASE IN HISTORY

Heralded as a scientific miracle, the pesticide Dichloro-diphenyl-trichloroethane (DDT) was used to control malaria during World War II. **By the mid 1950s, Americans were duped into allowing over 31 million acres of land to be sprayed with this "harmless" poison.** Years later congressional reports tell a totally different story.

Yet scientists of the day knew DDT stored in fatty tissue "...at least as early as 1945 that DDT presents a potential residue problem." [1] Stammers, et al. noted in 1947, "*The possibilities of cumulative effects from storage of DDT in milk and tissues of sheep and cattle require further investigation.*" [2] In the late 1940's, Schechter and colleagues developed methods to detect contamination, "*of such products as milk, butter, eggs, meat, and fats when farm animals consume DDT treated feed,*" and show that DDT, "*can be excreted in milk.*" [3]

Remember those dates, folks, because claiming ignorance that "we didn't know back then" doesn't quite work anymore. Moreover, it was hoped the DDT contamination problem

would "just go away" and those provisions were inserted in section 331(c) of the Economic Opportunity Act of 1964.

> *"Until recently there was reasonable prospect that the problem [pesticide residues] would have diminished to a point by June 30, 1967, that further extension of the authority would not be necessary. However, in recent months, severl large producers in the Rio Grande Valley have **had their milk removed from the market because of DDT residue.** The local dairymen, Texas and New Mexico State agencies, dairy and cotton associations, and our Agricultural Research Service, have undertaken to determine the causes by tests in Arizona."*

> SOURCE: *Dairy Indemnity Payments.* 90th Congress, 1st Session. Senate Report 476. August 3, 1967

It's pretty apparent the public never was supposed to know about that happening.

Got DDT?

When it was obvious pesticide residue levels weren't dropping as hoped, the Dairy Indemnity Payment Program, or DIPP, officially was created through public law in 1968 to reimburse farmers whose milk was contaminated with pesticides. However, in reading the congressional Dairy Indemnity Payments reports that talk candidly about the widespread contamination, it is then that we see the full extent of the cavalier agricultural practices promoted by the U.S. Government.

> *"The problem which our dairy farmers are facing has been brought about by the use of chemicals approved by the Federal Government to dust crops. Some of these chemicals have been found to contaminate feeds. The contamination passes on into the milk and when the residues of pesticides is [sic] found to be of too high a level, the farmers are forced to dump their milk, taking it out of commercial channels. The result has been disastrous to the dairy farmers involved, some of which have had to go into bankruptcy."*

> *"There had been at that time, reasonable prospect that the problem would have diminished to a point that further extension of the authority would not have been necessary.*

However, as the Department testified then, several large producers had their milk removed from the market because of DDT residue. The problem had in fact not been solved. The local dairymen, the State governments involved, dairy and cotton associations, and the U.S. Department of Agriculture have been cooperating in an effort to rid us of this problem. The problem, however, still continues and it appears that it will continue for the forseeable (sic) years to come."

SOURCE: *Dairy Indemnity Payments.* 90th Congress, 2nd Session. Senate Report 1363. July 8, 1968

What does DDT have to do with vaccines? Well, I'm glad you asked.

A Spoonful of MediSIN...

To prove the safety of DDT, apologists often point anecdotally to **Merck** Chemist, Joseph J. Jacobs, slurping spoonfuls of the white stuff before public speeches. This act, to them, apparently validates the chemical's safety across the board for all persons. Fortunately, Joseph wasn't lactating at the time.

Wait a minute! Did I just say Merck? Who do you think was a producer of DDT?

Not only did Merck mass-produce DDT for controlling malaria in Italy during WWII, **Merck also formulated and manufactured a polio vaccine in response to the 1952 polio epidemic. The detection of the cancer-causing virus (SV40) in polio vaccine by Merck scientist,** Maurice Hilleman, will be the focus of a future article.(http://en.wikipedia.org/wiki/Maurice_Hilleman)

Enter Henry Kumm.
Henry Kumm worked at the International Health Division of the Rockefeller Foundation for Medical Research in 1928. During World War II, Kumm experimented with larvicides containing DDT to control the spread of malaria in Italy. [4]

Additionally, according to the Medical Archives at John Hopkins, *"In 1951 [a year before the polio outbreak], he resigned from the Rockefeller Foundation to accept a position as assistant director of research at the National Foundation for Infantile Paralysis. He conducted field trials in the study of gamma globulin and the Salk vaccine and became the director of research in 1954. Rejoining the Rockefeller Foundation in 1959, Kumm retired as an associate professor in 1964." [5]*

To tie up a few loose ends, <u>you should know the National Foundation for Infantile Paralysis (NFIP) was supposed to investigate the cause of polio.</u> That foundation also had invested millions of dollars into researching a possible polio vaccine prior to the 1952 polio epidemic.

Years after the polio epidemic in 1979, researchers Gabliks and Utz reported, *"Studies in cell cultures with insecticides...indicated increased replication of poliovirus in human cells exposed to Kelthane, Karathane, and DDT. Furthermore an activation of a latent virus was also observed in primary rabbit kidney cells grown in the presence of DDT."* [6]

Do you think it would be a cold day in hell before Good Old Henry Kumm would make a possible correlation between a polio outbreak and the spraying of pesticides? **However, there were more profits to be made from polio vaccines than recommending people stop drinking DDT-contaminated milk, wasn't there?**

> *"The phenomenal growth in production and use of chemical control agents is illustrated by the fact that in 1940 these products had a wholesale value of about $40 million. Today (August 12th, 1959) it is $290 million and is estimated to reach $1 billion by 1975. One-sixth of all croplands and millions of acres of forests, rangelands, and marshlands are treated annually with these chemicals..."*

> *"...Before 1940, relatively small amounts of such chemicals as nicotine, rotenone, pyrethrum, and the aresenicals (sic) were used for insect control. During and following World War II a rapid changeover to DDT, heptachlor, dieldrin, TEPP, malathion, and related compounds occurred."*

> SOURCE: Authorizing Research on Insecticides, Herbicides, Fungicides, and Other Pesticides by the Secretary of the Interior. 85th Congress, 2nd Session. House of Representatives Report No. 2181. July 16, 1958.

In Conclusion

In order to keep the American people hoodwinked about the toxicity of DDT, attempts are still made by corporate interests to dispel the dangers. Yet, each new piece of information brings us shockingly closer to the truth. At the end of the line, the core message still remains stubbornly in-line with the corporate sickness care agenda: take more vaccines, consume more pills, and wind up getting sicker!

As a result of DDT application, can we just be imagining more cancer, more autism, and more chronic illnesses? Despite superficial attempts by government agencies to 'get to the bottom' of a problem, taxpayers shouldn't be suckered into waiting twenty years to get answers. I certainly don't want to wait at all, since toxic products should not be permitted into the food chain at any phase of production. P.T. Barnum said, *"There's a sucker born every minute,"* and that's what they apparently are relying on.

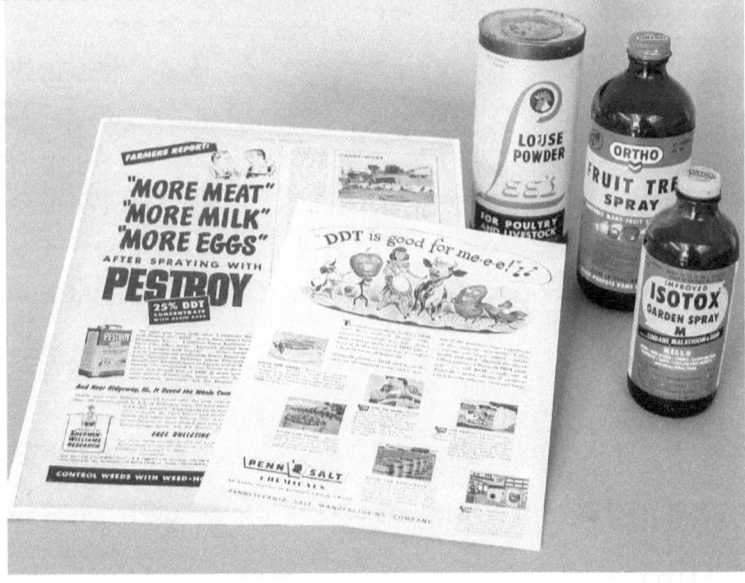

NO MATTER WHO APPROVES OR RECOMMENDS IT – SOME THINGS ARE WRONG AND DANGEROUS.

TABLE OF CONTENTS.

PART TWO

PART ONE

EXPECT REJECTION

'EXPERTS' & 'AUTHORITY'

PSYCHOLOGY of Non Believers

OPINIONATED MINDS

Incidental thoughts – fluoride, vaccine,
press articles

PLACEBO EFFECT – Cancer

New research – Chemo

Bruno, Galileo

A New Nomenclature

Socrates method - rebuttal

Yes, the truth will be ignored, and its proponents mocked. Be prepared.

THE FALLACY AND FRAUD OF 'AUTHORITY'.

The article below in *italics* is produced from my writings and appeared in my first major book, "I Can See Clearly Now", interlude between parts 2 and 3, published November 2011. Things and conditions have not changed since then, of course.

*"The absolute dangers to mankind in the use of **fluoride** in water supplies, **amalgam** fillings (loads of toxic mercury) in dental work, **vaccines**, **chemotherapy**, radiotherapy, electroshock treatments, **microwaves** (not only in stoves), and a host of similarly dangerous issues have been not only ignored, but actively and willfully suppressed. (If you doubt this, then do some Internet research to establish all issues and concerns.) **These and other issues are loudly and vehemently declared as quite safe for use or implementation by the "authorities". The media carries that biased one-sided opinion, the dictates of "authority" without question and gives little to no coverage of the nature and issues of concern.** This has been going on for decades; it has not changed and probably will never change. **Protest is virtually useless.** Why is this?*

It is as though there really is an orchestrated assault against mankind on as many fronts as can be found.

***Authority goes unquestioned in or by the media, allowing the most ludicrous of claims to become established as indisputable facts in the minds of the populations.** It is well known that if someone shouts the most obvious of lies loud and long enough it will come to be accepted by most.*

*Therein is a key to a lot of problems. A lot of people simply do not **think** for themselves, and are accepting of what "**authorities**" dictate. Think about this. **What is an expert or authority in most cases? Most are merely people who have gone through some educational institutions and learned what to accept, what to think, what to say, and having proven by examination that they can effectively recite the dogma of their branch of study, they are awarded some certification to that effect.***

3

These then become new peers to the closed group that monitor all aspirants and venturers into their field of "expertise".

[**Added 2 April 2014**: People forget or are unaware of a most interesting truth and fact regarding the fresh new breed of "experts", and "authorities". Once they have 'graduated' from their chosen field of academic 'expertise' and study, having efficiently learned the dogma and able to faultlessly recite it, **they must then find a job**. Now, where would one expect such new and enthusiastic graduates to gain employment? Probably within the very corporates or institutions that provided and probably funded their very education. Giving loyalty and service to the hands that fed and educated them is natural. Now, follow that line of thought. Are they going to disagree with or support any individual of group that is in total opposition to their now established dogma? The corporates, drug company's etc generously fund the education and research facilities expecting a perpetuation of their approval and support. Nursing schools and training sources fall unfortunately into the same category. See the essay herein *"Empty Caskets"*.]

Thus such a self-perpetuating closed group of "experts", custodians of the sacred dogma, <u>forever control any given branch of learning or endeavour. Thus they can ignore protest with impunity, for they and their gospel will outlive any given protesters.</u> A good example of this in action is the age of the **Sphinx** *issue. The evidence of geologists presented decades ago is still totally ignored by "***Egyptologists***", who shifted the discussion to how "insensitive" the protesters were. The real issues have never been addressed or answered, <u>as ignoring them has proven the most effective of manner in which to deal with such embarrassing (for them) questions or issues. This same tried and proven effective method is used with regular monotony whenever experts or authority (political leaders are skilled in its use) are faced with indisputable valid evidence of error in their thinking and dogma.</u>*

We see questions and error ignored in many fields. Theory of evolution is one dealt with in this book. "Big Bang" theory is quite nonsense and impossible to reasonably understand or accept. There are numerous cloudy issues we are asked to blindly accept because we are told to and quite simply that is the way things are. If "they" speak, we are expected to remain silent and accepting for the thinking has been done on our behalf. Religion is rife with "don't think about it" issues.

Why don't most people think?

We will look into that very basic question in a section that deals with the **education** *system imposed upon us by our diligent and vigilant government authority. We are carefully brought up and educated so as not to think independently. Our common* **religions** *train us in acceptance and not to think for ourselves. And should you join any* **military** *force, then in most cases and issues you become forbidden to think.*

The populations at large are mostly **compliant** *and subservient to the authority and experts that they enable and allow to rule and preside over them. We will look into this issue as well. We will see that there may indeed be "something in the water" that should give us all cause for major concern. Yes collectively we drink the* **reticulated water**, *eat the* **processed foods**, *have our* **vaccine** *shots, take the* **medications** *of corporates, then wonder why there is an epidemic of early deaths or the populations are obese and unhealthy. We are addicted to and enslaved by an epidemic of alcohol, tobacco, and addictive drugs. Organized crime is almost in control in some areas.*

Most **religions** *are far from user-friendly. We will look at this claim in more detail later and see that untold human misery and death result because of this insidious control method. Now understand that there is a huge difference between "religion" and "**spirituality**". I am not against the latter, but assuredly have no more time for the former. Essentially "spirituality" is what one has or feels from "within", it is an expression of self. "Religion" is from "without", extraneous to the self and is a set of proscribed scripture, doctrine, dogma, beliefs, creeds, etc. It is control and regulation. It is rarely conducive to peace and harmony. Also understand that I am not against "GOD", but that I have no time for and am against those named entities proclaimed as gods by the many "religions" or sects. "GOD", or the intelligent field of the universe has absolutely nothing whatsoever to do with religion. Yes I know we have all been brought up and trained quite differently than to even think that kind of thought."*

[The next article also in *italics* was written and published in my same book, Part 2. article 5.]

"3: Don't blindly believe the so-called experts or authorities. Their track record is not too good. People were killed once for disagreeing with their assertions that the earth was no only flat, but the center of all that there is or can be. Men like **Galileo** *often had to hide their discoveries. Do the research; doctors and the medical profession*

*actually cause the death of more people than from any other cause. Dentists still cram teeth with proven deleterious **amalgam** fillings containing some 40% mercury, one of the most dangerous of heavy metals. Then of course the health experts of the various ministries, the governments etc., all under guidance of the "experts", dump the most toxic of chemicals into our water supplies, I speak mainly of **fluoride**. Do some research on fluoride, for instance, do a google on "fluoride and Nazis". The issue of fluoride will be taken up in more detail later when we come to consider its true purpose in being force-fed to populations. The list of experts in error is almost endless.....*

Learn to ask an appropriate question, then seek the answer yourself."

So then, the above material is by way of introduction, and vital at that, to the material that follows.

"Science" is thus not an absolute or accurate construct.

We will see that "Science" is often used, can and is frequently used as a convenience to prostitute, add credibility to otherwise nefarious claims, to make the (gullible) public accept their claims as endorsed by the "experts" and acceptable "authorities".

Some of the claims almost make the eyes water in pain in retrospect, but this is the history of prostitution of "science" in the cause of profits for their proponents. We will see another major strategy is to use "famous" people, known "authorities" or "experts", and historically, 'celebrities' to endorse their claims and 'products'.

The purpose and intent of this book is to endorse what I said in two previous books. "It Ain't Necessarily So."

Finally, this book is NOT a sequential novel; you can read in any sequence, to jump to any subject. ENJOY.

The following article is off the Internet as an essay and comments from "Activist Post. A most remarkable essay indeed.

Sunday, January 19, 2014

The Psychology of Being a "Non-Conspiracy Theorist"

Bernie Suarez

Activist Post

There is a brand of people amongst us. They have no name but they exist. They are everywhere, at work, at home, at school, and in the streets, stores, and shopping malls. It is highly unlikely to not know someone who belongs in this category. It's the so-called non-conspiracy theorists. **You know, the guy who tries to terminate conversations by <u>alleging that you are nothing more than a "conspiracy theorist"</u> and the information you share is false or not believable.** Yes, that guy. Let's meet face to face with your typical non-conspiracy theorist. We all know them, they often are the ones who hold the "conspiracy" verbal accusation as a valid logically defined argument in and of itself. The logic works like this:

Conspiracy theorist says: "You are claiming that fire alone can cause a building to self-implode, descend at freefall speed into its own footprint? That's physically impossible, what about Newton's Laws and laws of thermal dynamics and such? "

Non-conspiracy theorist says: "No, you are wrong <u>because</u> you are a conspiracy theorist."

And with that, often the non-conspiracy theorist will walk away. What's happening? They had nothing to elaborate on, so the non-conspiracy theorist - whose thinking is engineered and controlled by government, mainstream media and Hollywood entertainment - resorted to a socially engineered answer. For this reason it is fascinating to explore the mindset and psychology of the individuals who take this position in place of a logical stance.

Core defect and twisted meanings

By definition, conspiracy means a group of two or more people secretly plotting (or conspiring) a harmful (or evil) deed against another person(s). This behavior is part of human nature. **Humans have been conspiring against each other since the beginning of time.** There was a never a time in world history when such an elementary behavior (of conspiring against an enemy) did not exist. Can you find a period in history when a certain human emotional trait or action didn't exist? Was there a time when the human genome didn't express jealousy, game playing, free trading, rage, or happiness? As odd or ridiculous as this may sound, non-conspiracists unknowingly subscribe to this logic. **If evidence points to a conspired crime, why not treat it as such? Why demonise the very concept of conspiring?** This is a core defect in the thought process of non-conspiracy theorists. You can easily identify them by their speech.

What the non-conspiracy theorist doesn't see is that the battlefield is right before them and **they have outsourced all critical thought and action to a government thinking service known as mass media.** Like being on the football field while the ball is in play, without realizing what is happening. You see people waving at you in the stands to get out of the way, but you don't understand what they are saying so you continue walking on the field with your headset on. Likewise, non-conspiracy theorists perceive all logic and reason arguments and warnings made by the conspiracy theorist as noise. They don't understand the warning and so they continue living and carrying on with what they are doing. Like the person on the field not understanding the warning being communicated by those in the stands, **non-conspiracy theorists cannot receive the basic signals of logic and reason.**

The headsets adding to the confusion is like the TV and mainstream media news. **Since they themselves <u>refuse to look at the evidence,</u> they put their faith in the government and in the mass corporate establishment to safely guide them in their reality.** They gain a sense of psychological protection from this overall system. Since the information being believed is almost always artificial they need to hear their own opinions repeated to them by the voices on TV so they can confirm (and re-confirm) their own belief system to themselves to be legitimate. At no point will the non-conspiracy theorist plan a day of research or dedicate a few hours every once in a while to research the topic or put any thought into issues.

Information and world problems are but one category in a shelf of categories that make up their lives. The non-conspiracy theorist ignores that government has always implemented social engineering and mass mind control on the general public. **In order for the non-conspiracy theorist to confidently walk away from someone <u>who challenges their belief system with scientific facts,</u> they need to have a sustained comfort and assurance that what government and mainstream media is saying is true.**

Maintaining the non-conspiracy delusion

The non-conspiracy theorist is thus profoundly psychologically interconnected with today's mass mind control paradigm. They are a species representing a full by-product of 21st-century social engineering. The doctrine of this type of social engineering programs its believers to believe that when government and media accuses someone of "conspiracies" then this accusation is cause for someone being considered diseased. The symptoms being paranoia. **But paranoia is based on systematized delusions and delusions are based on false beliefs.** A proper exploring of meanings brings us back to proving what is true or false. We come full circle and the spin is over. **Non-conspiracy theorists don't realize they dwell in this circle of <u>misapplied words</u>, never exploring the meanings or doing the work to determine what is true and what is provable.**

Non-conspiracy theorists therefore wake up every morning and reach for the mental orientation map known as mainstream media news. Without it they would be disoriented, as they would not know what to believe. <u>They actually believe that if anyone was guilty of wrongdoing at the highest level of government someone would speak out every time and everyone would know about it.</u> They ignore that state secrets are the norm and government operations are conducted in secret. <u>They ignore the consequences each individual at the government and military level faces for blowing the whistle.</u> Despite these consequences **many individuals at government level still risk it all and do blow the whistle on government. Despite all this, the idea of maintaining State secrecy is a myth to non-conspiracy theorists. The idea that government would do something immoral, nefarious or criminal is a fiction as well to the non-conspiracy theorist.**

The history of war, corruption, tyranny and fascism is incidental, coincidental, insignificant and irrelevant to the non-conspiracy theorist. None of these should be used to gauge the events of our times since history is a thing of the past. Non-conspiracy theorists choose not to connect the historical dots. According to them, there is nothing to learn from the history of tyranny and totalitarianism. **Anyone attempting to connect the dots is likely a conspiracy theorist.** This goes hand in hand with the logic defect we discussed earlier. As the earlier case of the conspiracy theorist having his scientific arguments debunked by virtue of simply being diseased with the accusation of conspiracy theorist. **Note, the non-conspiracy theorists use the name as a loaded, proven concept with power to permanently label someone diseased. In this case, attempting to connect the dots automatically tags you as a conspiracy theorist.**

As far as non-conspiracy theorists are concerned, all critical thought is deferred to the authorities. The scientists do their work and then report the facts to the government and mainstream media who carefully announce to the masses what they should know. **Any scientists who speak out and claim to have evidence contradicting government claims must first be endorsed and approved by the mainstream media and government.** Without this approval the scientists are marginalized no matter how large their numbers are. *Without these ridiculous rules the delusion of being a non-conspiracy theorists cannot be maintained.* **This is the psychology of tyranny**. *Tyranny and totalitarianism cannot be implemented without mass mind control. This dangerous group-think mentality will ignore all warning signs to help continue advancing the agenda. And so the non-conspiracy theorist is the most important instrument for maintaining control of the masses. Without these non-conspiracy theorist vessels of the global empire, the plan would not be possible.*

It is quite possible that years from now the concept of conspiracy will return back to where it belongs right next to other concepts like jealousy, laughter, love, stealing, fighting, friendship, hatred and other human expressions that define who we are. By then perhaps being a friendship theorist, a stealing theorist or a jealousy theorist will be the new propaganda bogey monster term. Or perhaps years from now we will be sick and tired of this vicious cycle of physical and mental slavery. Perhaps people will truly have enough of the control system and will have abolished it by then.

Standing by truth

I am thrilled to be considered a "conspiracy theorist" by those still controlled by the social engineering experiments of the last one hundred or so years. **We are proof that humans are able to critically think on their own and prefer to be free.** We are the embodiment of the resistance, we never run from challenges and debates, we recreated the media and are responsible for the progressive death of the mainstream media. We choose to give humanity a voice outside of the socially engineered control system. We rely on firsthand accounts, physical evidence consistent with natural laws, factual documents, common sense, high probability, and forensic scientific evidence before jumping to conclusions. We do not rely on name-calling to give strength to our position and we stand as a reminder that the human mind will never be completely contained.

Unravelling the reality we face became normal for us at some point and the information the non-conspiracy theorists consider scary is common news to those of us name called "conspiracy theorists".

In order to be considered a conspiracy theorist by the non-conspiracy theorists you must believe that government is corrupt. Believing that others are corrupt does not get you the title automatically. **The list of what qualifies someone as a conspiracy theorist has been changing rapidly and today is determined by mainstream media and government in real time. Today, only government and mainstream media gets to decide who qualifies as a conspiracy theorist.** The thought process of the non-conspiracy theorist is thus predictable and automated because the agenda being followed by government and mass media is predictable. **The mindset is: no research required if Associated Press, NBC, ABC, CBS, FOX, CNN or NPR doesn't agree to report it, then all other sources must be false.** <u>This shines a light into the thought process of non-conspiracy theorists.</u>

Impact on humanity

How creepy does all this sound? We often take for granted the thought process required to make someone believe how they believe and the (mainstream media believing) non-conspiracy theorist mentality is quietly as culpable for the condition of the world today as the individuals who actually carry out the crimes that have put us where we are. Little to no effort is put by non-conspiracy theorists to learn about their own social mindsets, social engineering, group-think, and government propaganda and mind-control history.

The non-conspiracy theorists are thus arguably indirectly ushering in many of the religious and cultural prophesies of the end times and the predictions of the doomsday prophets. They are proving to be key vessels in the events that are to come. They have already carved their mark in history as supporters of the global empire. Everyone who has played a role in supporting the global empire of the U.S. has already left their mark. Let's all wait and see how this mental defect will play itself out and what role these non-conspiracy theorists will play in ushering in the final pieces of the global government. Will they be rewarded? Will they support the extermination of all critical thinkers? Will it turn out the non-conspiracy theorists were expressing an alternate DNA or is this idea too far out? Will it turn out they were all part of a larger experiment? Will the final waking up process or critical mass be stomped out by this core of non-conspiracy theorists who blindly believe government-programmed lies? Or will they be responsible for delaying critical mass by a certain amount of time.

These and many other questions will be answered in the next few years/decades. Let us not forget the layers of defective logic and blind faith in government required to be considered a non-conspiracy theorist. It sounds bizarre but it's true. It's been said that in times of mass deceit that telling the truth is a revolutionary act; this has never been more apparent. Want to make an impact on others? Want to be a giant among men? Then tell the truth and watch the sheep run. There is so much deceit in today's world that if you blindly reverse everything government and mainstream media says, just by default you would be closer to truth than if you believed even some of what they say.

Remember the non-conspiracy theorist who says "someone would have spoken out"? We've all heard this excuse. Of course, someone always does speak out; only they call those who speak out, no matter how high in government they are, conspiracy theorists, instead of someone who is speaking out or blowing the whistle. It's time to memorialise the web of non-logic that qualifies a non-conspiracy theorist and not take for granted what these individuals mean to our battle for freedom and what key role they will play in the final lockdown of what was once a beacon for freedom throughout the world. Even as each of us carries on every day it's difficult to fathom how the average person you come across who is a non-conspiracy theorist is having such a massive impact on millions of people globally and the direction of humanity as a whole including the overall survival of the human race.

*Bernie Suarez is an activist, critical thinker, radio host, musician, M.D, Veteran, lover of freedom and the Constitution, and creator of the **Truth and Art TV** project. He also has a background in psychology and highly recommends that everyone watch a documentary titled The Century of the Self. Bernie has concluded that the way to defeat the New World Order is to truly be the change that you want to see. Manifesting the solution and putting truth into action is the very thing that will defeat the globalists.*

Anonymous said...
The cognitive barriers people erect against contemplation of conspiracies like 9/11 is something which fascinates me. Perhaps that's because I myself had to step through that barrier (around 2007), when I watched a video lecture by Dr. Steven Jones, which I found unassailable.

The author of the article is exactly right when he claims most people have outsourced their critical thinking. They rely on the mass media to define their worldview, and what concepts are kooky. The mass media succeeds largely because they project a convincingly consistent picture from different political and scientific angles. I also blame the public education for laying the ground work for this intellectual outsourcing. Those who ask uncomfortable questions in class are quickly normalized. :)

One argument against "conspiracy theories" I hear a lot is: "Sure, governments are riddled with corruption, but it's dog-eat-dog system. Everyone's out for themselves and **there's no way anyone could plan and execute such a plot. And even if they did, someone would blow the whistle."** *(Bob adds: Many people DO "blow the whistle". Then in almost all cases, they who whistle blow are called 'conspiracy theorists". Weird that.)*

Yet these same people have no difficulty believing that two billion years ago some single-celled organisms began working together to form staggeringly complex multi-cellular life forms--all without the least bit of conscious planning. It just happened. Why don't they apply the same evolutionary thinking to human organizations and cultural forms? Does the introduction of consciousness make this aggregation of purpose suddenly more difficult? I don't think so.

Not all life evolved the same way. Once speciation began, life took on many diverse forms. Personally **I think of the "ruling elites" (to be lazy and use a popular shorthand) as a kind of predator,** a societal bird of prey that circles above us in a sphere of human activity most people do not even think about. Hawks are not well adapted for life on the ground, but their detachment and vastly different method of movement makes them very good at finding prey, catching them, and eating them.

The general public hasn't yet learned to look up, and see the circling threat. But they will, eventually.
January 19, 2014 at 8:05 AM

Anonymous said...
...I am a "Federally Protected Whistle-Blower"...after having refused to go along with the type of corruption my incompetent bosses were engaged in...and after I was interviewed by OSI and told them the truth...my boss called me in to tell me how they would "get me"...my "career was over"...etc.,etc....but I tape recorded him...and "won" my case....for which they beat the crap out of me for two decades.. before I retired...

So the writer is correct...you have to understand how corrupt

our government is...in order to understand how the..."compartmentalized cowardice" works in government. Each employee is often given parts A,B & C to work on...knowing nothing of parts X,Y & Z...by the time they have learned that their part of the job involves some degree of illegality..they are 15 years into their cubicle..and do not want to risk the retirement and benefits...it is easier to simply go along...sad...but true...I am so glad I never lived or worked like that...besides..most corrupt people..are simply stupider and lazier that successful, hard working folks...

RJ O'Guillory
Author-
Webster Groves - The Life of an Insane Family
　　　January 19, 2014 at 11:18 AM

Paul said...
Galileo: 'I theorise that the world is not flat and that the Church is wrong in their teachings in this regard, I theorise that it is round'.

Joe Public: 'shut up, conspiracy theorist.'
　　　January 19, 2014 at 12:24 PM

Veri Tas said...
These people are lazy, fearful and deceitful (The People of the Lie).

Our society is structured so as to make sweeping the truth under the carpet pay; they promote the yes-men and women and demote the critical thinkers and the change agents.

It's pretty hard to expend so much energy on acting and speaking with integrity and not to be rewarded for it.

The blander and more zombied out the person - the more "well-adjusted" (either through self-censorship of the mind and speech or Prozac-induced) - the better the outcome for the person in society.

　　　January 19, 2014 at 6:30 PM

Tampa Dave said...

15

Excellent article! None of us who reads it will feel as intimidated by the catchphrase again.

"Those who ask uncomfortable questions in class are quickly normalized."

Lol. I read this "marginalized", **and believe this is the secret true goal of State Education: to marginalize us (which takes away our voice) so that we cannot be a threat to them. How many "free thinkers" can fascist societies support?**
         January 19, 2014 at 7:37 PM

The Ceej said...
Everything likes to think they're reasonable and rational, but I find none of us are. Even when we try our hardest to learn and think critically and have others attempt to find the flaws in our logic (yeah... good luck with that... you're likely to just get insulted instead), we still have our emotional biases... We may be better at reason and logic than those who choose not to even try, but we could be even better if we acknowledge how bad we are. I doubt we will ever get to the truth...

I take issue with the term, "conspiracy theory." It's a term that's literally meaningless. <u>Someone says, "That's a conspiracy theory," they really just said, "You're claiming that two or more people are collaborating to achieve a common goal." And? What's your point?</u> You really haven't added anything new to the conversation with that statement. And <u>the fact that it's used as a label to dismiss a statement outright is as absurd as it is offensive.</u> Imagine the following conversation:

<u>"Where are Dick and Jane?"
"They're not home. They must have gone out to buy a loaf of bread for the house."
"That's a conspiracy theory."</u>

<u>Yes. Yes, it is. But, does it make it absurd? Is it a reason to dismiss it outright? No. It's a perfectly plausible explanation based on the first viewing of the circumstantial evidence. Further research may indeed suggest that they DID go to the store to buy a loaf of bread.</u>

This last part of my comment is for Veri Tas, regarding your statement, "It's pretty hard to expend so much energy on acting and speaking with integrity and not to be rewarded for it."

But, you DO get rewarded for it... Just not in the immediate. Following the herd will keep you safe and stable, but with no substantial rewards. You may gain status or money, but you'll never have any real relationships. If you're lucky, you may be able to lie to yourself about the frivolous relationships you DO have. If you act with integrity and honesty, you get punished quite frequently for it in the immediate, but the long-term rewards (which I'm only now, about five years after I started trying to act with integrity and honesty) make it all worth it in the end. Although, if I were doing it for the rewards, I'd have given up a long time ago and have never known about them...

Jesus, I've made this almost as long as the original article. Sorry about that.
    January 20, 2014 at 12:24 AM

Sandy Lunoe said...
We were several in Norway who criticized the swine flu vaccine Pandemrix in 2009 and warned against increased risk of miscarriages and autoimmune condtions.

We were called conspiracy theorists by the Norwegian Inst. of Public Health and the media.

Tragically, our warnings materialised. When the many cases of narcolepsy (an autoimmune condition) rolled in the leaders of the Inst. of Public Health quietly left their positions.

We are still often called conspiracy theorists in discussions. People seem to like attaching labels to others.
    January 20, 2014 at 4:35 AM

Anonymous said...
I think you're dancing around what the real problem is. You say that the non-conspiracy theorist is immune to logic or critical thinking. But, the real issue is that most Americans (can't speak for the rest of the world) think in terms of courtroom logic, not real logic. In courtroom logic, the argument is discontinued the moment a flaw is discovered in

it. This means that if there is a chain of logic, from A->B->..Z, no matter how "true" Z might be, if there is any flaw in any of the steps from A through Y, then that is reason alone to discount Z. This means that in order to convince most people of something, you have to have an airtight case. You have to have all of the possible bases covered, and you must me unassailable in other more subtle facets of your argument like in your use of emotion, how complex your argument is, and your personal background.

In other words, in order to convince most people that a conspiracy exists then, even if it is blatantly obvious, you have to have a completely airtight argument, and you must present it in a way that is easy to understand, in a timespan that is not too short or too long, and you must "be" of a certain character that nobody can articulate, but they'll know when they see it. If you falter along any step of the way, you're done.

I'm not saying that's right, I'm just saying that that's the way it is.

This becomes particularly problematic when you realize that there is no such thing as objective truth. The degree to which things are true or false is mostly proportional to the degree to which we *agree* things are true or false. You can see this playing out in UFO reports. How many times do you see in a UFO abduction reports, "This happened to 500 people, including, 2 policemen (who we are supposed to believe fit that certain character), therefore it is classified as officially unexplained" vs. "This happened to 10 people. Nobody can corroborate their story. Clearly it was a hoax." vs. "David Icke thinks people are being abducted. He has a fairly compelling case up to where he goes completely off the rails and starts talking about reptilians. So even though we more or less believed him up until hour 10 of his 15 hour slide show, we're going to have to go and classify him as nuts. Even worse, because his unfortunate conclusion can't be proven, even though two thirds of his argument is based on sound evidence, we're afraid we have to disqualify him. Also, you may not rely on any part of his argument to build a new one. Case closed."

Now even worse, by definition of conspiracy and the fact that you are not in on it, you have an opposition that is actively exploiting the fact that people rely on courtroom logic and are looking for a reason to not believe your argument, using a variety of tactics that grows over time. You call it the "mainstream media", but let's be honest, it's the people in charge of the conspiracy USING the mainstream media as a tool which they control.

Is it any wonder that things are the way you have described them?
January 20, 2014 at 7:01 AM

Bob Williams said...

I often encounter another pyscho-sociall dynamic, especially amongst electricians and other skilled trades - they intentionally take the normative view with a full consciousness that the details of the official account of 9/11 do not add up. In other words, they choose not to be troublemakers. As one electrician told me - "I ask myself two questions, a) is there anything I can do about it? and b) am I going to do anything about it? If the answer is no to either question, I forget about it".

January 20, 2014 at 12:21 PM

Anonymous said...

The problem lies in the questions raised and not being able to give one pinpoint answer. We live in a society that wants answers NOW and has no patience (or maybe fear) in examining the possibilities, the vastness of possibilities out there.
Comfort in not knowing, believing what they are told? Comfort in government as good? Comfort in the home/job/family/friends? Maybe just fear on why the bell is ringing and how it can destroy what they think they know.
January 20, 2014 at 2:52 PM

This is another article off Internet Activist Post

Beware of the Opinionated Mind

Beverly Blanchard | *Ancient Wisdom* 532 VIEWS
December 16, 2013

They are out there. Those individuals who will zealously defend a particular belief even though the information presented before them proves it is wrong. Trapped in a fixed mode of thinking, these individuals will exasperate and frustrate you. **No matter how you present the information they will refuse to believe you.**

Usually after engaging in a conversation with these types of individuals, you will find yourself feeling as though someone just sucked the life force out of you. **Your energy will be depleted and often you will be emotionally charged with anger or resentment.** You may even begin to doubt your own beliefs.

There are some who have referred to these individuals as psychic vampires or energy drainers. In reality, however no one is sucking your energy. **On a vibrational level, it is you who is not maintaining your frequency. You are allowing someone else's energy to dominant the interaction and in doing so you have gone down to their level.** So just how do you maintain your frequency?

There are three primary ways we can view a situation. The first is through your own eyes. The second is through the eyes of the other. Finally there is the director which is really you as the independent observer. **If you find that you are getting emotionally dragged into a conversation or situation, it is time to move to the independent observer position.** *To do this you remove your focus from the conversation and mentally step outside your body. You become like a fly on the wall. Usually when* *you do this you discover the conversation is not worth pursuing or you will see how you are allowing your energy to be manipulated.*

There are numerous strategies people use to unconsciously manipulate a conversation or interaction. One of the most prevalent techniques is that of not listening. If you are dealing with an opinionated mind, they will automatically shut down their physical senses. The first sense to be turned off is sound. They may be physically in front of you but they aren't hearing a word you are saying. In fact most are occupied in internal dialogue within their mind and are filtering everything you are saying through their own beliefs and perceptions.

If the information you are presenting does not fit into their tightly enclosed box of beliefs, many times they will start editing the conversation. Usually when this happens you will find they will rewrite what you are saying and often the interpretation does not even match what you are actually saying. At this point, you will find yourself continually saying, "I didn't say that."

Another strategy that the opinionated mind likes to use is "put-downs". Whatever you say, they will make into a joke or they will mock you. *In this case, these strategies are based on fear. The points you are making are endangering their belief systems and the only way they can protect themselves is by cutting you down.*

If you find yourself engaged in a conversation with someone which is going nowhere, the best thing you can do is disengage. You can change the subject or simply walk away. To try to convert someone to your way of thinking is futile and will only disrupt your energy. Of course, if you like arguing you can remain in the interaction but then you have to ask yourself: What is this person reflecting back to me about my own opinionated mind?

INCIDENTAL THOUGHTS. (As compiled and written by me, "Now & Then", 2nd edition Sept 2012.)

FLUORIDE

<u>The resistance to removing fluoride from our water and the insistence of using it indicates more is involved than mere dental concerns.</u> **Claims for the efficacy of fluoride in the improvement of dental hygiene have <u>long ago been seriously and credibly discredited</u>, but this is disregarded.** It is telling to see the strategy used by proponents, and the language used to decry opponents as "hysterical", ignorant, unlearned, ("conspiracy theorists") etc. But who really is hysterical and resorts to bluster, name-calling, and an assumed status of "authority" and academic superiority?

If we put aside the heated bluster for a moment and seek a motive for any possible and claimed conspiratorial motive, we do find reasonable and credible motives. Is that coincidence? To me the very fact that the "argument" is so heated and emotional in itself indicates more at stake than the mere resolution of a disagreement. Were it mere disagreement based on interpretation of fact alone, this could be resolved very fast and efficiently by open discussion in some very public forum, where indisputable fact could be established and the errant party allowed to resolve their differences.

Were this allowed (by whom?) then we would find the debilitating and long term effect of fluoride poisoning (No, that is not "hysterical" use of a word, but a known fact, fluoride is extremely toxic.) on the populations. Then a true conspiracy is revealed, as fact not theory.

<u>But the danger is vehemently denied with only bluster and authoritive claims.</u> No new inquiry or fair and balanced research open to public scrutiny is allowed or contemplated by the authority that dictates its use upon an acquiescent public. Why is this?

We are led to an inescapable conclusion: The conspiracy is very real and <u>we should be very afraid because really, more is at stake and involved than the single issue of fluoridating or poisoning the public water supply.</u> A lot more is involved.

Probably there never will be a response even to the ethics question posed by dictatorial mass medication and doping of the entire citizenry. Its time we impose our individual embargo on drinking dosed water, and using it only to flush down the sewers, where most of it goes anyway.

As an afterthought question, it is interesting to wonder how or where, if this toxic waste were not disposed of by its creators through the public kidneys via our reticulated water supplies, would this vast amount of waste be effectively and safely disposed? It is right up there with nuclear waste, but with the additional disadvantage of having no "half-life". **Not only would vast fortunes by lost in income, but also expended in safe disposal. Yes a lot is at stake indeed.**

VACCINES

I was looking at a very young infant today. If it were a cat it's eyes would not be open yet. The horrible thought crossed my mind that its parents would probably put it through the vaccination cycle, dosing it with almost endless inputs of dangerous and toxic compounds, and this when it is most vulnerable. (21 vaccines in 1st 5 months of life!)

We need to simply look behind the curtains. We need to ask some simple but important questions. **What is our "health service" and who are the "authorities"?**

We will very soon discover it is not driven by a benevolent group to ensure the well-being of either our children or us, but by multinational corporations controlling the entire medical system for profit and power, with the manipulation of unimaginable wealth and power.

But doctors sustain it and recommend it? Maybe, but doctors also recommended one smoke Camels or another brand of cigarette a few decades ago. They recommended the use of leeches or bleeding one for the cure of illness. There is an almost endless list of sheer rubbish and error of opinion by medical authority. **Right now most do not recommend the use of simple supplements of vitamins or minerals for the well-being of an individual.** Cancer? Best to cut, burn, or poison it, as they assure us there are no alternatives. Yeah right.

Again more is involved than one may think or imagine. Again, it not so then why is there so much bluster, name calling, character assignations, indeed the same strategies used as listed above for use of fluoride?

It is self-evident when one looks that there is **active massive and global suppression and repression of not only opposition to vaccinations, but also to cover up of the deaths and disasters directly caused by it.** Some details in my other books.

There is enough evidence "out there" to effectively dictate that many, if not all, vaccination immunizations etc should be immediately stopped pending open, honest, **INDEPENDENT** investigation determine permanent cessation, and compensation awarded to a vast number of injured and family of those killed by the practice. (Hereunder are merely two New Zealand cases that somehow got into the press.) The publication of these two articles really surprised me; as such events are generally almost entirely suppressed.

GOD AND DOCTORS. What is the difference between God and a doctor? God does not think he is a doctor.

ACC in legal battle over $250,000 bill from flu jab

Kiran Chug

dompost.co.nz

A WOMAN who fell seriously ill after having a flu jab has been left with a medical bill of more than $250,000, which the Accident Compensation Corporation is refusing to pay.

Allison Cottle is embroiled in a legal battle with ACC as she prepares a High Court challenge to its decision not to pay, which a judge says is legally sound.

Mrs Cottle, who has permanent New Zealand residency but lives in Cleveland, in the United States, contracted a life-threatening illness a month after having the flu jab in March 2007.

A week after the vaccination, she was on holiday in the US, where she was diagnosed with Guillain-Barre syndrome, which causes spinal paralysis, after suffering a suspected stroke.

Yesterday, she told *The Dominion Post* of the unfairness of the decision made by ACC and the impact it was having.

"It's very stressful. My husband is unwell and we're just working through the matter ourselves."

Mrs Cottle did not want to comment further while the matter was before the courts.

Her lawyer, John Miller, said appeal papers had been filed in the High Court against the earlier decision made by Judge Beattie.

"It's your nightmare scenario of getting ill in the US and having to have an emergency treatment."

Court documents show Mrs Cottle went on holiday to the US a week after receiving the flu jab known as Vaxigrip in March.

After suffering a suspected stroke, she was taken to hospital, where the seriousness of her condition saw her admitted to intensive care and not given medical clearance to return home until July.

She was faced with a medical bill of US$180,482 (NZ$250,000), which ACC said it would not pay.

When Mrs Cottle asked for the decision to be reviewed, ACC was ordered to pay for the treatment.

However, it has now successfully challenged that review through the courts, with Judge Beattie finding the law was on ACC's side as the legislation stated that it should not pay for acute treatment received overseas. In his judgment, Judge Beattie said Mrs Cottle could not have known the flu jab would cause her to develop Guillain-Barre syndrome when she left for her holiday in the US, or that it would have such a "deleterious effect" on her well-being.

Although, legally, ACC must not pay for treatment received overseas, Judge Beattie said the case exposed a possible gap in the legislation.

In his decision, he described Mrs Cottle's situation as a "classic example" of one where ACC should have discretionary power to pay for her treatment.

A spokesman for ACC said the judge had found that ACC had interpreted the law correctly when making its decision.

Mr Miller said he had previously represented a client who developed a similar syndrome to Mrs Cottle after having a flu jab.

That client won her case for ACC to cover her treatment as she had developed the illness in New Zealand.

He was unsure whether Mrs Cottle had travel insurance when she went on holiday.

From New Zealand, Wellington Dominion Post, 10 Sept 2010.

The following article same paper, 9 Apl 2011.

Vaccine-mixup toddler has cancer

Natalie Akoorie

A HAMILTON toddler, who was mistakenly injected with a vaccine to prevent cervical cancer when he was just six weeks old, has developed a rare form of leukaemia.

Chace Topperwien began chemotherapy for the M7 strand of acute myeloid leukaemia, on his second birthday last month, but doctors cannot tell whether the Gardasil vaccine he was injected with in May 2009 caused the cancer.

His parents, Ryan and Keri Topperwien, are devastated at the diagnosis after they were reassured at the time of the incident that their baby would not suffer any adverse reactions.

"I had thought in the very back of my mind that the absolute worst thing they could say is leukaemia," Mrs Topperwien said. "When they said that he had it, it blindsided both of us."

The couple, both 27, were horrified when, at six weeks old, their son was given the vaccination meant for teenage girls instead of one to prevent meningitis. Gardasil targets the human papillomavirus, which is responsible for 99 per cent of cervical cancers.

Since the cancer diagnosis, Chace has been undergoing an aggressive course of chemotherapy treatment, for up to 10 hours a day, in Starship children's hospital in Auckland. The once energetic and outdoor-loving preschooler is confined to his bedroom 23 hours a day while his immune system recovers – a simple cold or bug could be disastrous.

The couple said their only child is taking the illness in his stride and other than a nasogastric tube for feeding, looks deceivingly well.

"He's still smiling and laughing. Other than not being allowed out of his room, he's still happy. But he knows that he's not well."

Meanwhile, they want their concerns about a possible link between the Gardasil vaccination mistake and Chace's leukaemia acknowledged. "It's not a crazy theory," Mrs Topperwien said.

Waikato District Health Board medical officer of health Felicity Dumble said she did not believe there was a link between Gardasil and leukaemia.

"I'm not aware of leukaemia ever being caused by Gardasil or any other vaccine."

Waikato Times

Student's meningococcal death 'isolated case'

Kate Newton
HEALTH

HEALTH officials say the death of a Wellington student from meningococcal disease is an isolated case.

precaution, Dr McKenzie said.

A Wellington Hospital spokeswoman said that would include flatmates but probably not other students with whom the woman had shared classes.

Regional Public Health was waiting for results that would show which strain of the disease the woman had, or whether she had been vaccinated.

The bug is passed through close contact, such as sharing drinks, living in the

familycare
PRIMARY HEALTH ORGANISATION

FREE
CHILDHOOD IMMUNISATION

Why should your child be vaccinated?

- Immunisation can protect your child from serious diseases, which can cause serious complications, including death
- Childhood vaccines on the National Immunisation schedule are free, safe and effective
- Immunisation is based on sound scientific evidence
- Uses the body's natural defence mechanism, the immune response, to build resistance and protect your child from specific infections

For more information see your Doctor or Practice Nurse

BE WISE - IMMUNISE

These scandalously inaccurate ads appear regularly in local press, (This one Upper Hutt Leader 27 Apl 2011) and stress "safe and

effective" etc. It is an obvious lie, evidenced by the innumerable reports, as sampled in this book, and such ads the result either of oblivious ignorance, or willful disregard for publicly known and reported facts. I point out actual Ministry of Health publications DO admit there is some risk associated with vaccinations, but the notification is so low key as to be almost impossible to readily see. I believe the placement of such misrepresenting ads to be criminal.

OUR HEALTH AUTHORITIES and ministry frankly are not to be trusted nor relied upon for beneficial advice or service. After all, they make false claims, endorsements, or statements. Such as:

Fluoride is safe to ingest and not dangerous.
Amalgam fillings with mercury are OK to use.
We should not use or need dietary supplements.
There are no health risks or dangers from cell phone use.
Vaccinations are safe and effective.
Cell Phone transmission towers present no health risks.
The only treatment for cancer is chemo or radiation therapy.
Fluoride reduces dental caries.
Aspartame is approved as safe in foods and drinks.
Discredit and malign natural remedy, Homeopathy etc.
Approve of a chemical cocktail of toxic food additives.

THE PLACEBO EFFECT CLAIM.

This week I saw a fascinating report of a T.V. current affairs programme. It involved an aging couple (over age 60) and their life-style. She had cancer that had now totally disappeared from her body. The report then dealt with their assertion as to the causes for this cure.

They exercised extensively and ran literally hundreds of marathons each year. Exercise established, then they focused on diet and foods. They ate no meats at all, ate only fruit, vegetables nuts and the like, but added to that regimen was the fact that they ate only raw foods. Absolutely no cooked foods at all. Yes they did indeed appear to be in excellent health, looked very physically fit, and very mentally active and alert. Their entire demeanor was most impressive, such as to make one consider taking up a similar life style for self.

Now we should all know and be aware that it is common knowledge that a "correct" diet and elimination of less desirable foods is not only good for ones health, but totally recommended by wise and knowledgeable health authorities. One sees posters on hospital walls and doctors surgeries everywhere teaching the necessity for good food and diet. Adverts frequently appear on TV also admonishing good diet and correct supplements. All good so far, and the report on the show are very correct and acceptable.

The reporter then introduced an "expert and authority" apparently in all matters connected with the subjects of report.

Stunningly, she smugly made the statement that "there is <u>no evidence or scientific proof</u>" that such things as diet (etc.) had any effect and certainly would not cure cancers. Compounding her smirking (literally) adverse comments she declared any 'seeming' improvement would only be the result of the "placebo effect."

I was instantly impressed that her attitude, smugness, and ignorance was totally in line with the pro-fluoride and pro-vaccine "errorists".

In an attempt to discredit any claims for healing by natural methods, homeopathy and the like, the allegation is made that besides <u>being only anecdotal evidence and unscientific, any claimed or evident healing would be ascribed to the "placebo effect"</u>.

That is a marvelous claim to make, and it makes me wonder about an "about turn" for such an allegation. For example, **<u>hereunder are results of chemotherapy treatment for various cancers treated in over 80,000 Australian adults.</u>** The percentage of effective results (5 year survivors) is stunningly low. **Surely this would evidence to most open-minded people that it must surely therefore be an unscientific method of healing.** Besides that, the numbers enabled to survive 5 years or more would fall very far short of any commonly accepted rates of efficacy due to the "placebo" effect. **We could well claim that any recovery due to chemotherapy must therefore surely be solely a result of the placebo or belief effect, <u>and at the same time discern something very wrong with the "treatment" to give results so far below expected placebo effect healings.</u>**

To illustrate my comment, about half way down the list we see 10,661 persons of which only 1.5%, 164 people survive 5 years!

Received: 18 August 2003 Revised: 20 April 2004 Accepted: ...

CYTOTOXIC CHEMOTHERAPY SURVIVAL IN ADULT MALIGNANCIES

Table 1 – Impact of cytotoxic chemotherapy on 5-year survival in Australian adults

Malignancy	ICD-9	Number of cancers in people aged >20 years[a]	Absolute number of 5-year survivors due to chemotherapy[b]	Percentage 5-year survivors due to chemotherapy[b]
Head and neck	140–149, 160, 161	2486	63	2.5
Oesophagus	150	1003	54	4.8
Stomach	151	1904	13	0.7
Colon	153	7243	128	1.8
Rectum	154	4036	218	5.4
Pancreas	157	1728	–	–
Lung	162	7792	118	1.5
Soft tissue sarcoma	171	665	–	–
Melanoma of skin	172	7811	–	–
Breast	174	10661	164	1.5
Uterus	179 + 182	1399	–	–
Cervix	180	867	104	12
Ovary	183	1207	105	8.7
Prostate	185	9869	–	–
Testis	186	529	221	41.8
Bladder	188	2802	–	–
Kidney	189	2176	–	–
Brain	191	1116	55	4.9
Unknown primary site	195–199	3161	–	–
Non-Hodgkin's lymphoma	200 + 202	3145	331	10.5
Hodgkin's disease	201	341	122	35.8
Multiple myeloma	203	1023	–	–
Total		72964[d]	1696	2.3%

*Numbers from Ref [21].
[a]Absolute numbers (see text).
[b]% for individual malignancy.
[d]Total for Australia 1998 = 80,864 people.

New study finds that <u>CHEMO</u> targets cancer cells that were dying anyway

Tony Issacs

FW: Jan. 29, 2014

(Originally published ov. 7, 2011)

(NaturalNews) Researchers in a new study have found that cancer cells best targeted by chemo are already on the verge of self-destruction. In the study, published online by the journal *Science* on October 27, researchers found that cancer cells that are closer to the threshold for programmed death via apoptosis are more susceptible to chemotherapy than other cancer cells that have yet to reach that stage.

"Many chemotherapy agents work by damaging structures within cancer cells, particularly DNA and microtubules [tiny tubes used for a variety of cell functions]," reported the study's senior author, Anthony Letai, MD, PhD, of the Dana-Farber Cancer Institute which conducted the study. "When the damage becomes so extensive it can't be repaired, the cells initiate a process known as apoptosis, in which they sacrifice themselves to avoid passing the damage on to their descendants."

The researchers reported that they had found a way to profile cancers according to how close the cancer cells were to death, and this could help determine which ones might be most susceptible to chemo drugs. However, the study also appears to show that oftentimes chemo at best merely speeds the process of cellular death, which every healthy cell is programmed to do. The study thus raises some important questions about the use of chemotherapy.

For example, would those cancer cells that respond to chemotherapy have died naturally anyway? And, if chemotherapy mostly only hastens a natural process, is it worth damaging healthy cells and organs and going through the debilitating effects of deadly chemo drugs?

As Natural News has often reported, **chemo does not cure <u>cancer</u> to begin with, and often chemo spreads cancer or**

causes damage (which kills the patient before the cancer can). See, for example:

Chemo Does Not Cure: Often It Inflicts Damage and Spreads Cancer

http://www.naturalnews.com/027028_cancer_WHO...

Studies have shown that at best chemo results in only a 2-3% increase in survival time. Chemo does nothing to address the root causes of cancer: it merely kills some cancer cells (as well as healthy ones). Neither does chemo do anything to prevent the return of cancer. In fact, the damage done by chemo to the natural immune system and the rest of the body makes it more likely that either the original cancer or another form of cancer will return.

Chemo is part of the three-legged failed mainstream cancer treatment paradigm, which tries to address cancer by merely poisoning out, cutting out and/or burning out the mere symptoms of cancer. Of the three, chemo is by far the most commonly used option. It is also the most lucrative.

The average oncologist practice derives 75% of its income from the mark-ups it makes on chemo drugs, which it buys at wholesale prices and then heavily marks up to sell to the patients who are prescribed the drugs. Notably, in one survey 75% of oncologists reported that they would not use chemo or prescribe it for their family members due to the low success rates and horrific side effects. Yet, 75% of cancer patients are prescribed chemo.

Which leads to yet another question: Is the use of chemo really about healing? Perhaps the answer can be found in this quote by outspoken chemotherapy opponent and oncologist, Glen Warner, M.D.:

"Chemotherapy is an incredibly lucrative business for doctors, hospitals, and pharmaceutical companies ..The medical establishment wants everyone to follow the same exact protocol. They don't want to see the chemotherapy industry go under, and that's the number one obstacle to any progress in oncology."

See also:

Hiding the Truth About Losing the War on Cancer

http://www.naturalnews.com/023286_cancer_can...

Dying Cancer Patients are Milked for Every Last Dollar

http://www.naturalnews.com/029032_cancer_pat...

Other sources included:

http://www.wddty.com/chemo-targets-cancer-ce...

http://www.dana-farber.org/Newsroom/News-Rel...

http://www.cancer.gov/newscenter/pressreleas...

About the author

Tony Isaacs, is a natural health author, advocate and researcher who hosts The Best Years in Life website for those who wish to avoid prescription drugs and mainstream managed illness and live longer, healthier and happier lives naturally. Mr. Isaacs is the author of books and articles about natural health, longevity and beating cancer including "Cancer's Natural Enemy" and is working on a major book project due to be published later this year. He is also a contributing author for the worldwide advocacy group "S.A N.E.Vax. Inc" which endeavors to uncover the truth about HPV vaccine dangers.
Mr. Isaacs is currently residing in scenic East Texas and frequently commutes to the even more scenic Texas hill country near Austin and San Antonio to give lectures and health seminars. He also hosts the CureZone "Ask Tony Isaacs - featuring Luella May" forum as well as the Yahoo Health Group "Oleander Soup" and he serves as a consultant to the "Utopia Silver Supplement Company".
http://www.naturalnews.com/z034075_chemotherapy_cancer_cells.html

Here are historical records of "Authorities" gone hopelessly wrong and in error. With tragic results.

giordano bruno – **Giordano Bruno, philosopher and scientist, burnt at the stake 400 years ago**

By Frank Gaglioti
16 February 2000

Four centuries ago today, on February 16, 1600, the Roman Catholic Church executed Giordano Bruno, Italian philosopher and scientist, for the crime of heresy. He was taken from his cell in the early hours of the morning to the Piazza dei Fiori in Rome and burnt alive at the stake. To the last, the Church authorities were fearful of the ideas of a man who was known throughout Europe as a bold and brilliant thinker. In a peculiar twist to the gruesome affair, the executioners were ordered to tie his tongue so that he would be unable to address those gathered.

Throughout his life Bruno championed the Copernican system of astronomy which placed the sun, not the Earth, at the centre of the solar system. He opposed the stultifying authority of the Church and refused to recant his philosophical beliefs throughout his eight years of imprisonment by the Venetian and Roman Inquisitions. His life stands as a testimony to the drive for knowledge and truth that marked the astonishing period of history known as the Renaissance—from which so much in modern art, thought and science derives.

In 1992, after 12 years of deliberations, the Roman Catholic Church grudgingly admitted that Galileo Galilei had been right in supporting the theories of Copernicus. The Holy Inquisition had forced an aged Galileo to recant his ideas under threat of torture in 1633. But no such admission has been made in the case of Bruno. His writings are still on the Vatican's list of forbidden texts.

An Infinite Universe

The universe was infinite for Bruno. There is infinite number of worlds with infinite number of beings living on them. Each world is finite within the infinite universe. The infinite universe is immobile because there is nothing beyond it, nowhere to move to. The unity of the universe is stable in its oneness and so remains forever. This is the basic characteristic of the universe that unites everything within itself. Beyond the multiplicity of forms and things, there lies the principle of unity, which is a key to understanding the similarities among different things such as between men and animals, men and plants, our planet earth and other planets, our sun and the other suns. With this unity, the Greek philosophers could declare a basic principle of ancient philosophy;

A Living Universe

He described a living universe, which we find in Plato as well, a universe like an enormous animal where infinite solar systems are like the cells of an enormous body. In this universe, all things are alive, as all parts of our bodies alive. The celestial bodies were the great living creatures for Bruno as it was in the philosophy of Pythagoras and Plato. These celestial beings were more than living beings. They are divinities such as the deities Mars, Jupiter, Venus, Mercury and Apollo, which are associated with the planets and the sun. In this living body, there is a continuous process of corruption and generation of forms. Bruno explains corruption and generation of forms as same because they derive from the same principle. The end of a corruption is the beginning of the generation of another thing so corruption is a generation and generation is a corruption.

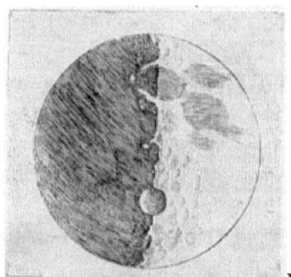

Galileo (1564-1642) was the first astronomer to make full use of the telescope, observing the craters of the moon and the satellites of Jupiter. His open advocacy of Copernican cosmology led, however, to a clash with the Catholic Counter-Reformation, and he spent his final years under house arrest.

Inquisition and first judgement, 1616

Pope Paul V (1552-1621), who ordered that the inquisitorial commission's 1616 judgement be delivered to Galileo by Cardinal Bellarmine.

On February 19, 1616, the Inquisition asked a commission of theologians, known as qualifiers, about the propositions of the heliocentric view of the universe.[28] Historians of the Galileo affair have offered different accounts of why the matter was referred to the qualifiers at this time. Beretta points out that the Inquisition had taken a deposition from Gianozzi Attavanti in November, 1615,[29] as part of its investigation into the denunciations of Galileo by Lorini and Caccini. In this deposition, Attavanti confirmed that Galileo had advocated the Copernican doctrines of a stationary Sun and a mobile Earth, and as a consequence the Tribunal of the Inquisition would have eventually needed to determine the theological status of those doctrines. It is however possible, as surmised by the Tuscan ambassador, Piero Guiccardini, in a letter to the Grand Duke,[30] that the actual referral may have been precipitated by Galileo's aggressive campaign to prevent the condemnation of Copernicanism.[31]

Deliberation

On February 19, 1616, the Inquisition asked a commission of theologians, known as qualifiers, about the propositions of the heliocentric view of the universe.[28] Historians of the Galileo affair have offered different accounts of why the matter was referred to the qualifiers at this time. Beretta points out that the Inquisition had taken a deposition from Gianozzi Attavanti in November, 1615,[29] as part of its investigation into the denunciations of Galileo by Lorini and Caccini. In this deposition, Attavanti confirmed that Galileo had advocated the Copernican doctrines of a stationary Sun and a mobile Earth, and as a consequence the Tribunal of the Inquisition would have eventually needed to determine the theological status of those doctrines. It is however possible, as surmised by the Tuscan ambassador, Piero Guiccardini, in a letter to the Grand Duke,[30] that the actual referral may have been precipitated by Galileo's aggressive campaign to prevent the condemnation of Copernicanism.[31] The _Index Librorum Prohibitorum_, a list of books banned by the Catholic Church. Following the Inquisition's 1616 judgment, the works of Copernicus, Galileo, Kepler and others advocating heliocentrism were banned.

Sentence

On February 24 the Qualifiers delivered their unanimous report: the idea that the Sun is stationary is "foolish and absurd in philosophy, and formally heretical since it explicitly contradicts in many places the sense of Holy Scripture..."; while the Earth's movement "receives the same judgement in philosophy and ... in regard to theological truth it is at least erroneous in faith."

At a meeting of the cardinals of the Inquisition on the following day, Pope Paul V instructed Bellarmine to deliver this result to Galileo, and to order him to abandon the Copernican opinions; should Galileo resist the decree, stronger action would be taken. On February 26, Galileo was called to Bellarmine's residence and ordered,"to abstain completely from teaching or defending this doctrine and opinion or from discussing it... to abandoncompletely... the opinion that the sun stands still at the center of the world and the earth moves, and henceforth not to hold, teach, or defend it in any way whatever, either orally or in writing." -The Inquisition's injunction against Galileo, 1616.

The Index Librorum Prohibitorum, a list of books banned by the Catholic Church. Following the Inquisition's 1616 judgment, the works of Copernicus, Galileo, Kepler and others advocating heliocentrism were banned.

With no attractive alternatives, Galileo accepted the orders delivered, even sterner than those recommended by the Pope.[32][33] Galileo met again with Bellarmine, apparently on friendly terms; and on March 11 he met with the Pope, who assured him that he was safe from persecution so long as he, the Pope, should live. Nonetheless, Galileo's friends Sagredo and Castelli reported that there were rumors that Galileo had been forced to recant and do penance. To protect his good name, Galileo requested a letter from Bellarmine stating the truth of the matter. This letter assumed great importance in 1633, as did the question whether Galileo had been ordered not to "hold or defend" Copernican ideas (which would have allowed their hypothetical treatment) or not to teach them in any way. If the Inquisition had issued the order not to teach heliocentrism at all, it would have been ignoring Bellarmine's position.

In the end, Galileo did not persuade the Church to stay out of the controversy, but instead saw heliocentrism formally declared false. It was consequently termed heretical by the Qualifiers, since it contradicted the literal meaning of the Scriptures, though this position was not binding on the Church.

Copernican books banned

Following the Inquisition's injunction against Galileo, the papal Master of the Sacred Palace ordered that Foscarini's *Letter* be banned, and Copernicus' *De revolutionibus* suspended until corrected. The papal Congregation of the Index preferred a stricter prohibition, and so with the Pope's approval, on March 5 the Congregation banned all books advocating the Copernican system, which it called "the false Pythagorean doctrine, altogether contrary to Holy Scripture."[32]

Francesco Ingoli, a consultor to the Holy Office, recommended that *De revolutionibus* be amended rather than banned due to its utility for calendrics. In 1618 the Congregation of the Index accepted his recommendation, and published their decision two years later, allowing a corrected version of Copernicus' book to be used. The uncorrected *De revolutionibus* remained on the Index of banned books until 1758.[34]

Galileo's works advocating Copernicanism were therefore banned, and his sentence prohibited him from "teaching, defending... or discussing" Copernicanism. In Germany, Kepler's works were also banned by the papal order.[35]

Trial and second judgment, 1633

With the loss of many of his defenders in Rome because of *Dialogue Concerning the Two Chief World Systems*, Galileo was ordered to stand trial on suspicion of heresy in 1633, "for holding as true the false doctrine taught by some that the sun is the center of the world", against the 1616 condemnation, since "it was decided at the Holy Congregation [...] on 25 Feb 1616 that [...] the Holy Office would give you an injunction to abandon this doctrine, not to teach it to others, not to defend it, and not to treat of it; and that if you did not acquiesce in this injunction, you should be imprisoned".[40]

With the loss of many of his defenders in Rome because of *Dialogue Concerning the Two Chief World Systems*, **Galileo was ordered to stand trial on suspicion of heresy in 1633, "for holding as true the false doctrine taught by some that the sun is the center of the world",** against the 1616 condemnation, since "it was decided at the Holy Congregation [...] on 25 Feb 1616 that [...] the Holy Office would give you an injunction to abandon this doctrine, not to teach it to others, not to defend it, and not to treat of it; and that if you did not acquiesce in this injunction, you should be imprisoned".[40]

Galileo was interrogated while threatened with physical torture.[35] A panel of theologians, consisting of <u>Melchior Inchofer</u>, <u>Agostino Oreggi</u> and <u>Zaccaria Pasqualigo</u>, reported on the *Dialogue*. Their opinions were strongly argued in favour of the view that the *Dialogue* taught the Copernican theory.[41]
Galileo was found guilty, and the sentence of the Inquisition, issued on 22 June 1633,[42] was in three essential parts:

- Galileo was found "vehemently suspect of heresy," namely of having held the opinions that the Sun lies motionless at the center of the universe, that the Earth is not at its centre and moves, and that one may hold and defend an opinion as probable after it has been declared contrary to Holy Scripture. He was required to <u>"abjure, curse, and detest"</u> those opinions.[43]

-

- He was sentenced to formal imprisonment at the pleasure of the Inquisition.[44] On the following day this was commuted to house arrest, which he remained under for the rest of his life.

- His offending *Dialogue* was banned; and in an action not announced at the trial, publication of any of his works was forbidden, including any he might write in the future.[45]

··

Do not think for one moment that Galileo was lacking fortitude in his seeming submission after a visit to the dungeon and torture threat. He was 36 years old

when Bruno was burned alive in Rome. Galileo knew the dangers only too well. Probably thought he was even lucky to still be alive in view of his writings, discoveries and public renown.

The articles above tell me there is a vast difference between the so-called "experts" and "authorities" and a very real "Elite".

Whereas the first two are in effect "stooges" and carry out or implement the will, desires, and ambitions of the latter, the elite are they who have the wealth in sufficient amount to be able to afford (pay) to do whatever they want, or get others to do it for them. It demonstrates the fact here are the "haves" and "have-not's", with one frequently willing and ignorant stooges for the "haves".

And with that in mind, I will create a new category and become a nomenclator.

Whereas "they" call many who actively protest "conspiracy theorists" henceforth such are "conspiracy identifiers".

And "those" who are not "conspiracy identifiers" must by default be "errorists" They may not necessarily be active supports of the errors, as such would surely now be "errorist activists". Thus we may identify "fluoride errorists", "vaccine errorists" etc.

I think that becomes a better and more appropriate title or classification to describe attitudes and states of ignorance or otherwise.

To further advance the nomenclature, we could well identify some among the "elite" who are really conspiracy "initiators", or "implementers". Enjoy the words.

On a slightly different tangent of this whole debacle, it behooves us to be able to find a satisfactory method to communicate with those ignorant and rude enough to refer to the Identifiers" as *conspiracy theorists*. *Ideally a response or reply that will not put us "on the back foot"*

<u>If one is to start off with all 'guns' blazing and make effort to show the words are both incorrect and insulting, you have just bought into a discussion (argument will follow) of at least two levels that you have almost no show of "winning", nor convincing of errors manifest by 1: their statement and 2: the subject under judgment.</u> Yes a minimum of two stages to cover to reach parity. Hopeless, and frustrating.

But how to achieve vindication?

This represents a fairly probable likeness of the Greek philosopher Socrates, 469 BC to 399 BC.

(Like Bruno and Galileo in above articles he also fell foul to the "Elite" of Athens, and was forced into 'suicide'. Lucky Columbus lived, probable because at sea and removed from 'elite' he had chance and time to prove his correctness.)

Hereunder lies the key to sorting out what could end up nasty and/or messy.

It was only the last week I remember that I learned all this well, and preaciced it extensively over a sales and management career that began in 1968.

Frankly, all discussion and 'argument' MUST become a matter of interview and questioning. It will become obvious.

Think about the following questions after you read my comments now.

They have just delivered to you what they know is a dismissal and obvious put-down. They are looking forward to progression with an attempt to belittle you, your thinking, and opinion. ("Oh you're wrong, 'every body' knows that.... And to think otherwise is just wrong and stupid." With implied or stated "Are you stupid?")

In some places them is just plain fighting words. And almost assuredly your blood pressure just went up, and you are now "on stage", disadvantaged.

<u>**DO NOT BIGHT, OR MAKE EFFORT TO ANSWER AT ALL.**</u>

Try to gain eye contact and pause for between 5 and 10 seconds. YES MENTALLYCOUNT THOSE SECONDS. Chances are he will give you at least 5 seconds before he may butt in again. So 5 seconds allow, then <u>clearly ASK YOUR FIRST QUESTION,</u>

Immediately he will be thinking along the lines of "no argument, are they interested in my thinking? Etc. Here come the question styles. I like the underlined questions.

<u>Conceptual clarification questions</u>

Get them to think more about what exactly they are asking or thinking about. Prove the concepts behind their argument. Use basic 'tell me more' questions that get them to go deeper.

- *Why are you saying that?*
- *What exactly does this mean?*
- *How does this relate to what we have been talking about?*
- *What is the nature of ...?*
- *What do we already know about this?*
- *Can you give me an example?*
- *Are you saying ... or ... ?*
- *Can you rephrase that, please?*

Probing assumptions

Probing their assumptions makes them think about the presuppositions and unquestioned beliefs on which they are founding their argument. This is shaking the bedrock and should get them really going!

- *What else could we assume?*
- *You seem to be assuming ... ?*
- *How did you choose those assumptions?*
- *Please explain why/how ... ?*
- *How can you verify or disprove that assumption?*
- *What would happen if ... ?*
- *Do you agree or disagree with ... ?*

Probing rationale, reasons and evidence

When they give a rationale for their arguments, dig into that reasoning rather than assuming it is a given. People often use un-thought-through or weakly-understood supports for their arguments.

- *Why is that happening?*
- *How do you know this?*
- *Show me ... ?*
- *Can you give me an example of that?*
- *What do you think causes ... ?*
- *What is the nature of this?*
- *Are these reasons good enough?*
- *Would it stand up in court?*
- *How might it be refuted?*
- *How can I be sure of what you are saying?*
- *Why is ... happening?*
- *Why? (keep asking it -- you'll never get past a few times)*

- *What evidence is there to support what you are saying?*
- *On what authority are you basing your argument?*

Questioning viewpoints and perspectives

Most arguments are given from a particular position. So attack the position. Show that there are other, equally valid, viewpoints.

- *Another way of looking at this is ..., does this seem reasonable?*
- *What alternative ways of looking at this are there?*
- *Why it is ... necessary?*
- *Who benefits from this?*
- *What is the difference between... and...?*
- *Why is it better than ...?*
- *What are the strengths and weaknesses of...?*
- *How are ... and ... similar?*
- *What would ... say about it?*
- *What if you compared ... and ... ?*
- *How could you look another way at this?*

Probe implications and consequences

The argument that they give may have logical implications that can be forecast. Do these make sense? Are they desirable?

- *Then what would happen?*
- *What are the consequences of that assumption?*
- *How could ... be used to ... ?*
- *What are the implications of ... ?*
- *How does ... affect ... ?*
- *How does ... fit with what we learned before?*
- *Why is ... important?*
- *What is the best ... ? Why?*

Questions about the question

And you can also get reflexive about the whole thing, turning the question in on itself. Use their attack against themselves. Bounce the ball back into their court, etc.

- *What was the point of asking that question?*
- *Why do you think I asked this question?*
- *Am I making sense? Why not?*

- *What else might I ask?*
- *What does that mean?*

Conceptual clarification questions

Get them to think more about what exactly they are asking or thinking about. Prove the concepts behind their argument. Use basic 'tell me more' questions that get them to go deeper.

- *Why are you saying that?*
- *What exactly does this mean?*
- *How does this relate to what we have been talking about?*
- *What is the nature of ...?*
- *What do we already know about this?*
- *Can you give me an example?*
- *Are you saying ... or ... ?*
- *Can you rephrase that, please?*

Probing assumptions

Probing their assumptions makes them think about the presuppositions and unquestioned beliefs on which they are founding their argument. This is shaking the bedrock and should get them really going!

- *What else could we assume?*
- *You seem to be assuming ... ?*
- *How did you choose those assumptions?*
- *Please explain why/how ... ?*
- *How can you verify or disprove that assumption?*
- *What would happen if ... ?*
- *Do you agree or disagree with ... ?*

Probing rationale, reasons and evidence

When they give a rationale for their arguments, dig into that reasoning rather than assuming it is a given. People often use un-thought-through or weakly-understood supports for their arguments.

- *Why is that happening?*
- *How do you know this?*
- *Show me ... ?*

- *Can you give me an example of that?*
- *What do you think causes ... ?*
- *What is the nature of this?*
- *Are these reasons good enough?*
- *Would it stand up in court?*
- *How might it be refuted?*
- *How can I be sure of what you are saying?*
- *Why is ... happening?*
- *Why? (keep asking it -- you'll never get past a few times)*
- *What evidence is there to support what you are saying?*
- *On what authority are you basing your argument?*

Questioning viewpoints and perspectives

Most arguments are given from a particular position. So attack the position. Show that there are other, equally valid, viewpoints.

- *Another way of looking at this is ..., does this seem reasonable?*
- *What alternative ways of looking at this are there?*
- *Why it is ... necessary?*
- *Who benefits from this?*
- *What is the difference between... and...?*
- *Why is it better than ...?*
- *What are the strengths and weaknesses of...?*
- *How are ... and ... similar?*
- *What would ... say about it?*
- *What if you compared ... and ... ?*
- *How could you look another way at this?*

Probe implications and consequences

The argument that they give may have logical implications that can be forecast. Do these make sense? Are they desirable?

- *Then what would happen?*
- *What are the consequences of that assumption?*
- *How could ... be used to ... ?*
- *What are the implications of ... ?*
- *How does ... affect ... ?*
- *How does ... fit with what we learned before?*
- *Why is ... important?*
- *What is the best ... ? Why?*

Questions about the question

And you can also get reflexive about the whole thing, turning the question in on itself. Use their attack against themselves. Bounce the ball back into their court, etc.

- *What was the point of asking that question?*
- *Why do you think I asked this question?*
- *Am I making sense? Why not?*
- *What does that mean?*

From memory this technique was known as or is in fact a "conversational rebuttal". And if used correctly in a true conversational tone, not as an interrogation or challenge, it will achieve more than simply drawing your guns to blaze away with the mouth.

THE DENTAL DISASTER

OR, CAN YOU REALLY TRUST THE DENTAL "PROFESSION"?

pictorial essay of past disasters

root canals and it's disaster

Amalgum toxic Mercury fillings

Distrust of dentistry

Fluoride supported by them

The following pictures are actual advertisements supported by the dental association and professionals decades ago. It evidences their support of "anything" that provided them income. Later it was obviously withdrawn and 'denied'.

DENTISTS VOTE OLD GOLD FIRST FOR THROAT-EASE

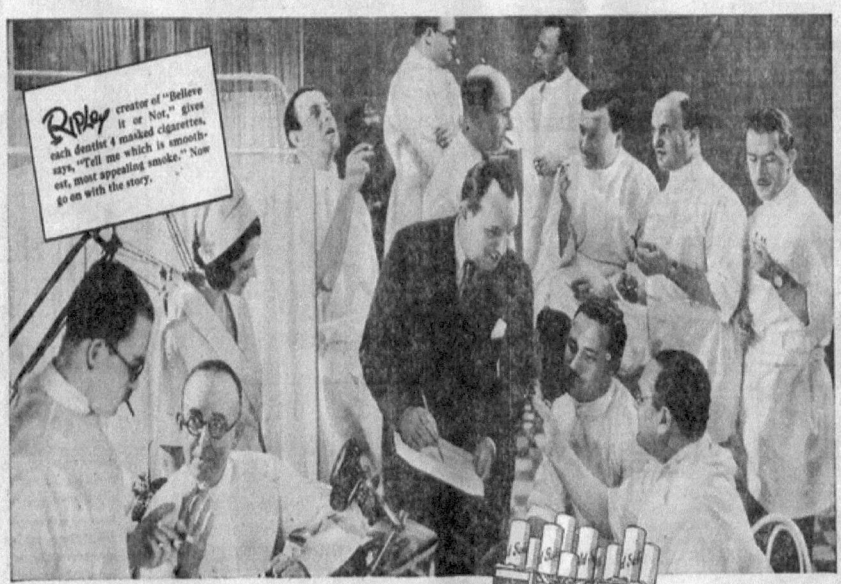

RIPLEY creator of "Believe it or Not," gives each dentist 4 masked cigarettes, says, "Tell me which is smoothest, most appealing smoke." Now go on with the story.

"BELIEVE IT OR NOT, this jury of dentists decided that OLD GOLD'S filling was perfect! They examined, smelled and smoked O. Gs.' better and smoother tobaccos. And the three rival cigarette brands limped home a poor second, third and fourth. Dentists see a lot of throats. Know what harsh tobacco does to teeth and gums. Know honey-smoothness when they meet it. Know clean, sun-ripened, *queen-leaf* tobacco when they taste it . . . masked or unmasked. Ask your dentist about OLD GOLD for throat-ease."

(Signed) ROBERT RIPLEY,
CREATOR OF "BELIEVE IT OR NOT!"

Old Gold CIGARETTES

THE TREASURE OF THEM ALL

© P. Lorillard Co., Inc.

CERTIFIED PROOF!
"I hereby certify that the following is the correct score of the cigarette test conducted by Robert Ripley, among Dentists, OLD GOLD 11; Brand X 5; Brand Y 4; Brand Z 4."

(Signed) J. S. M. GOODLOE
Certified Public Accountant
203 Broadway, New York

Not a Cough in a Carload

Tune in on OLD GOLD Character Readings . . . Tuesdays at 8:15 P. M., Thursdays at 9:15 P. M., Eastern Standard Time . . . Coast-to-Coast Columbia Network

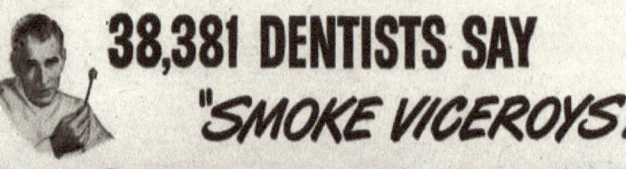

38,381 DENTISTS SAY
"SMOKE VICEROYS!"

VICEROYS FILTER THE SMOKE!

THE NICOTINE AND TARS TRAPPED* BY THE **VICEROY** FILTER *CAN NEVER STAIN YOUR TEETH!*

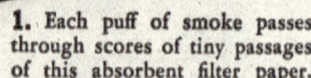

1. Each puff of smoke passes through scores of tiny passages of this absorbent filter paper.

2. The nicotine and tars thus trapped cannot stain your teeth— smoke is cooler, cleaner.

3. No tobacco crumbs can get in your mouth.

4. This filter is *exclusive* as is Viceroy's luxurious blend of fine domestic and imported tobaccos. Get Viceroys at your dealer's. You'll be glad you did.

*No filter can remove all nicotine and tars, nor does Viceroy make this claim.

Should a gentleman offer a Tiparillo to a dental hygienist?

"The doctor is a little late, sir. Will you have a seat?"

She's the best thing to hit dentistry since novocaine. "Hey Dummy," your mind says to you, "why didn't you have this toothache sooner?"

Maybe if…well, you could offer her a Tiparillo.® Or a Tiparillo M with menthol. An elegant, tipped cigar. Slim. And your offer would be cleverly psychological. (If she's a bit of a kook, she'll take it. If not, she'll be flattered that you *thought* she was a bit of a kook.) And who knows? Your next visit might be a house call.

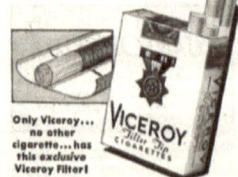

The hidden dangers of root canals you don't know about

Elizabeth Renter
Natural Society
Thu, 08 Nov 2012 11:29 CST

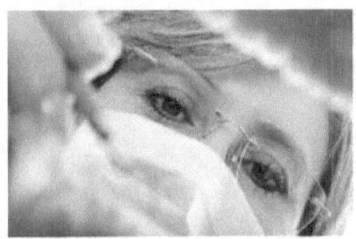

© Natural Society

Your dentist doesn't want you to know, and the American Dental Association (ADA) sure doesn't want us to tell you of the many dangers of root canals. After all, it's a multi-billion dollar industry. Any tainting of the root canal image could cost them serious cash, so (like Big Pharma) they deny there's any problems at all. Unfortunately for them, more and more people are being awakened to the trouble with traditional medicine and dentistry, so their industry *will* be taking a hit.

The Alliance for Natural Health says there are more than 25 million root canals performed in the United States each year, with 41,000 being performed every single day. And the number of dentists who discuss the true dangers of these procedures before they do them can probably be counted on one hand.

So, what is wrong with a root canal?

A root canal essentially removes the live pulp from a tooth and replaces it with a synthetic material. This stops the tooth from appearing to rot away, it does away with the internal damage that could be causing a toothache, the damage from an untreated cavity. But, while your dentist would have you think the root canal solves your problems - it really isn't that simple.

In addition to the central root of the tooth, where the dentist removes the tissue during a root canal, there are thousands of tiny side canals that aren't touched by your doctor. When the root is removed, the nerves in these side canals die. They rot. They fester and become a breeding ground for bacteria and infection.

Research has proven this to be the case. Not surprisingly, **the ADA denies the validity of any such research, maintaining that root canals are safe but refusing to scientifically refute the evidence to the contrary.**

Dr. Weston A. Price literally wrote the book on root canal dangers way back in 1922. His work was pooh-poohed by the ADA back then too. Since then, others have substantiated his work, which showed that **root canals are linked to immune diseases including Lou Gehrig's disease (ALS) and multiple sclerosis (MS).**

This disease is fueled by anaerobic bacteria surviving in the remaining roots of the teeth, in those tiny side canals. *(These are the tiny canals we see in the ads of T.V. for "sensitive Teeth" toothpaste Bob M..)* These bacteria don't need oxygen to survive and are quite happy living off of dead tissue. But this infection can and does spread, often without us knowing.

One study from Dr. Price **showed that a root-canaled tooth, when taken out from the human patient and placed in an animal, actually removed diseases like rheumatoid arthritis or heart disease from its host and followed the tooth into the animal. In other words, the human was cured and the animal with the root-canaled tooth developed the disease.**

In an Eastern medicine perspective, root canals also disrupt a link between certain teeth and body parts, since specific teeth actually lie on the same meridian point as various organs, tissues, glands, etc. Here is an interesting chart provided by Natural Dentistry.

What can you do? Look for alternatives. Namely, if your tooth is rotting, have it extracted. Then discuss natural options like a bridge or a titanium or zirconium implant. The Holistic Dental Association can help lead you to a holistic practitioner in your area.

97% of Terminal Cancer Patients Previously Had This Dental Procedure...

Visit the Mercola Video Library

Do you have a chronic degenerative disease? If so, have you been told, "It's all in your head?"

Well, that might not be that far from the truth... the root cause of your illness may be *in your mouth*.

There is a common dental procedure that nearly every dentist will tell you is completely safe, despite the fact that scientists have been warning of its dangers for more than 100 years.

Every day in the United States alone, 41,000 of these dental procedures are performed on patients who believe they are safely and permanently fixing their problem.

What is this dental procedure?

The root canal.

More than 25 million root canals are performed every year in this country.

Root-canaled teeth are essentially "dead" teeth that can become silent incubators for highly toxic anaerobic bacteria that can, under certain conditions, make their way into your bloodstream to cause a number of serious medical conditions—many not appearing until decades later.

Most of these toxic teeth feel and look fine for many years, which make their role in systemic disease even harder to trace back.

Sadly, the vast majority of dentists are oblivious to the serious potential health risks they are exposing their patients to, risks that persist for the rest of their patients' lives. The American Dental Association claims root canals have been proven safe, but they have NO published data or actual research to substantiate this claim.

Fortunately, I had some early mentors like Dr. Tom Stone and Dr. Douglas Cook, who educated me on this issue nearly 20 years ago. Were it not for a brilliant pioneering dentist who, more than a century ago, made the connection between root-canaled teeth and disease, this underlying cause of disease may have remained hidden to this day. The dentist's name was Weston Price—regarded by many as the greatest dentist of all time.

Weston A. Price: World's Greatest Dentist

Most dentists would be doing an enormous service to public health if they familiarized themselves with the work of Dr. Weston Price[j]. Unfortunately, his work continues to be discounted and suppressed by medical and dental professionals alike.

Dr. Price was a dentist and researcher who traveled the world to study the teeth, bones, and diets of native populations living without the "benefit" of modern food. Around the year 1900, Price had been treating persistent root canal infections and became suspicious that root-canaled teeth always remained infected, in spite of treatments. Then one day, he recommended to a woman, wheelchair bound for six years, to have her root canal tooth extracted, even though it appeared to be fine.

She agreed, so he extracted her tooth and then implanted it under the skin of a rabbit. The rabbit amazingly developed the same crippling arthritis as the woman and died from the infection 10 days later. But the woman, now free of the toxic tooth, immediately recovered from her arthritis and could now walk without even the assistance of a cane.

Price discovered that it's mechanically impossible to sterilize a root-canaled (e.g. root-filled) tooth.

He then went on to show that many chronic degenerative diseases originate from root-filled teeth—the most frequent being heart and circulatory diseases. He actually found 16 different causative bacterial agents for these conditions. But there were also strong correlations between root-filled teeth and diseases of the joints, brain and nervous system. Dr. Price went on to write two groundbreaking books in 1922 detailing his research into the link between dental pathology and chronic illness. Unfortunately, his work was deliberately buried for 70 years, until finally one endodontist named George Meinig recognized the importance of Price's work and sought to expose the truth.

Dr. Meinig Advances the Work of Dr. Price

Dr. Meinig, a native of Chicago, was a captain in the U.S. Army during World War II before moving to Hollywood to become a dentist for the stars. He eventually became one of the founding members of the American Association of Endodontists (root canal specialists).

In the 1990s, he spent 18 months immersed in Dr. Price's research. In June of 1993, Dr. Meinig published the book *Root Canal Cover-Up*, which continues to be the most comprehensive reference on this topic today. You can order your copy directly from the Price-Pottenger Foundation[ii].

What Dentists Don't Know About the Anatomy of Your Teeth

Your teeth are made of the hardest substances in your body.

In the middle of each tooth is the pulp chamber, a soft living inner structure that houses blood vessels and nerves. Surrounding the pulp chamber is the dentin, which is made of living cells that secrete a hard mineral substance. The outermost and hardest layer of your tooth is the white enamel, which encases the dentin.

The roots of each tooth descend into your jawbone and are held in place by the periodontal ligament. In dental school, dentists are taught that each tooth has one to four major canals. However, there are accessory canals that are never mentioned. *Literally miles of them!*

Just as your body has large blood vessels that branch down into very small capillaries, each of your teeth has a maze of very tiny tubules that, if stretched out, would extend for three miles. Weston Price identified as many as 75 separate accessory canals in a single central incisor (front tooth). For a more detailed explanation, refer to an article by Hal Huggins, DDS, MS, on the Weston A. Price Foundation website.[iii] (These images are borrowed from the Huggins article.)

Microscopic organisms regularly move in and around these tubules, like gophers in underground tunnels.

When a dentist performs a root canal, he or she hollows out the tooth, then fills the hollow chamber with a substance (called guttapercha), which cuts off the tooth from its blood supply, so fluid can no longer circulate through the tooth. But the maze of tiny tubules remains. And bacteria, cut off from their food supply, hide out in these tunnels where they are remarkably safe from antibiotics and your own body's immune defenses.

The Root Cause of Much Disease

Under the stresses of oxygen and nutrient deprivation, these formerly friendly organisms morph into stronger, more virulent anaerobes that produce a variety of potent toxins. What were once ordinary, friendly oral bacteria mutate into highly toxic pathogens lurking in the tubules of the dead tooth, just awaiting an opportunity to spread.

No amount of sterilization has been found effective in reaching these tubules—and just about every single root-canaled tooth has been found colonized by these bacteria, especially around the apex and in the periodontal ligament. Oftentimes, the infection extends down into the jawbone where it creates cavitations—areas of necrotic tissue in the jawbone itself.

Cavitations are areas of unhealed bone, often accompanied by pockets of infected tissue and gangrene. Sometimes they form after a tooth extraction (such as a wisdom tooth extraction), but they can also follow a root canal. According to Weston Price Foundation, in the records of 5,000 surgical cavitation cleanings, only two were found healed.

And all of this occurs with few, if any, accompanying symptoms. So you may have an abscessed dead tooth and not know it. This focal infection in the immediate area of the root-canaled tooth is bad enough, but the damage doesn't stop there.

Root Canals Can Lead to Heart, Kidney, Bone, and Brain Disease

As long as your immune system remains strong, any bacteria that stray away from the infected tooth are captured and destroyed. But once your immune system is weakened by something like an accident or illness or other trauma, your immune system may be unable to keep the infection in check.

These bacteria can migrate out into surrounding tissues by hitching a ride into your blood stream, where they are transported to new locations to set up camp. The new location can be any organ or gland or tissue.

Dr. Price was able to transfer diseases harbored by humans to rabbits, by implanting fragments of root-canaled teeth, as mentioned above. He found that root canal fragments from a person who had suffered a heart attack, when implanted into a rabbit, would cause a heart attack in the rabbit within a few weeks.

He discovered he could transfer heart disease to the rabbit 100 percent of the time! Other diseases were more than 80 percent transferable by this method. Nearly every chronic degenerative disease has been linked with root canals, including:

- Heart disease
- Kidney disease
- Arthritis, joint, and rheumatic diseases
- Neurological diseases (including ALS and MS)
- Autoimmune diseases (Lupus and more)

There may also be a cancer connection. Dr. Robert Jones, a researcher of therelationship between root canals and breast cancer, found an extremely high correlation between root canals and breast cancer.[iv] He claims to have found the following correlations in a five-year study of 300 breast cancer cases:

- 93 percent of women with breast cancer had root canals
- 7 percent had other oral pathology
- Tumors, in the majority of cases, occurred on the same side of the body as the root canal(s) or other oral pathology

Dr. Jones claims that toxins from the bacteria in an infected tooth or jawbone are able to inhibit the proteins that suppress tumor development. A German physician reported similar findings. Dr. Josef Issels reported that, in his 40 years of treating "terminal" cancer patients, 97 percent of his cancer patients had root canals. If these physicians are correct, the cure for cancer may be as simple as having a tooth pulled, then rebuilding your immune system.

Good Bugs Gone Bad

How are these mutant oral bacteria connected with heart disease or arthritis? The ADA and the AAE claim it's a "myth" that the bacteria found in and around root-canaled teeth can cause disease[v]. But they base that on the misguided assumption that the bacteria in these diseased teeth are the SAME as normal bacteria in your mouth—and that's clearly not the case.

Today, bacteria can be identified using DNA analysis, whether they're dead or alive, from their telltale DNA signatures.

In a continuation of Dr. Price's work, the Toxic Element Research Foundation (TERF) used DNA analysis to examine root-canaled teeth, and they found bacterial contamination in *100 percent of the samples tested.* They identified 42 different species of anaerobic bacteria in 43 root canal samples. In cavitations, 67 different bacteria were identified among the 85 samples tested, with individual samples housing between 19 to 53 types of bacteria each. The bacteria they found included the following types:

- *Capnocytophagaochracea*[vi]
- *Fusobacteriumnucleatum*[vii]
- *Gemellamorbillorum* [viii]
- *Leptotrichiabuccalis*
- *Porphyromonasgingivalis* [ix]

Are these just benign, ordinary mouth bugs? Absolutely not. Four can affect your heart, three can affect your nerves, two can affect your kidneys, two can affect your brain, and one can infect your sinus cavities... so they are anything BUT friendly! (If you want see just how unfriendly they can be, I invite you to investigate the footnotes.)

Approximately 400 percent more bacteria were found in the blood *surrounding* the root canal tooth than were found in the tooth itself, suggesting the tooth is the incubatorand the periodontal ligament is the food supply. The bone surrounding root-canaled teeth was found even HIGHER in bacterial count... not surprising, since bone is virtual buffet of bacterial nutrients.

Since When is Leaving A Dead Body Part IN Your Body a Good Idea?

There is no other medical procedure that involves allowing a dead body part to remain in your body. When your appendix dies, it's removed. If you get frostbite or gangrene on a finger or toe, it is amputated. If a baby dies in utero, the body typically initiates a miscarriage.

Your immune system doesn't care for dead substances, and just the presence of dead tissue can cause your system to launch an attack, which is another reason to avoid root canals—they leave behind a dead tooth.

Infection, plus the autoimmune rejection reaction, causes more bacteria to collect around the dead tissue. In the case of a root canal, bacteria are given the opportunity to flush into your blood stream every time you bite down.

Why Dentists Cling to the Belief Root Canals are Safe

The ADA rejects Dr. Price's evidence, claiming root canals are safe, yet they offer no published data or actual research to substantiate their claim. **American Heart Association recommends a dose of antibiotics before many routine dental procedures to prevent infective endocarditis (IE) if you have certain heart conditions that predispose you to this type of infection.**

So, on the one hand, the ADA acknowledges oral bacteria can make their way *from your mouth to your heart and cause a life-threatening infection.*

But at the same time, the industry vehemently denies any possibility that these same bacteria—toxic strains KNOWN to be pathogenic to humans—can hide out in your dead root-canaled tooth to be released into your blood stream every time you chew, where they can damage your health in a multitude of ways.

Is this really that large of a leap? **Could there be another reason so many dentists, as well as the ADA and the AAE, refuse to admit root canals are dangerous? Well, yes, as a matter of fact, there is. <u>Root canals are the most profitable procedure in dentistry.</u>**[x]

<u>What You Need to Know to AVOID a Root Canal</u>

I strongly recommend never getting a root canal. Risking your health to preserve a tooth simply doesn't make sense. Unfortunately, there are many people who've already have one. If you have, you should seriously consider having the tooth removed, even if it looks and feels fine. Remember, as soon as your immune system is compromised, your risk of developing a serious medical problem increases—and assaults on your immune system are far too frequent in today's world.

If you have a tooth removed, there are a few options available to you.

1. Partial denture: This is a removable denture, often just called a "partial." It's the simplest and least expensive option.
2. Bridge: This is a more permanent fixture resembling a real tooth but is a bit more involved and expensive to build.
3. Implant: This is a permanent artificial tooth, typically titanium, implanted in your gums and jaw. There are some problems with these due to reactions to the metals used. Zirconium is a newer implant material that shows promise for fewer complications.

But just pulling the tooth and inserting some sort of artificial replacement isn't enough.

Dentists are taught to remove the tooth but leave your periodontal ligament. But as you now know, this ligament can serve as a breeding ground for deadly bacteria. Most experts who've studied this recommend removing the ligament, along with one millimeter of the bony socket, in order to drastically reduce your risk of developing an infection from the bacterially infected tissues left behind.

I strongly recommend consulting a biological dentist because they are uniquely trained to do these extractions properly and safely, as well as being adept at removing mercury fillings, if necessary. Their approach to dental care is far more holistic and considers the impact on your entire body—not JUST your mouth.

If you need to find a biological dentist in your area, I recommend visiting toxicteeth.org[xi], a resource sponsored by Consumers for Dental Choice. This organization, championed by Charlie Brown, is a highly reputable organization that has fought to protect and educate consumers so that they can make better-informed decisions about their dental care. The organization also heads up the Campaign for Mercury-Free Dentistry.

References:

- [i] Weston A. Price Foundation
- [ii] Price-Pottenger Foundation

- [iii] Weston A. Price Foundation June 25, 2010
- [iv] Quantum Cancer Management
- [v] American Association of Endodontists
- [vi] Journal of Clinical Microbiology February 2007
- [vii] Journal of Clinical Microbiology July 2003
- [viii] Clinical Infectious Diseases June 1996
- [ix] Science Daily January 4, 2011
- [x] The Wealthy Dentist July 12, 2011
- [xi] ToxicTeeth.org

You are here: Home » Breaking News » **The Dangerous Truth About Root Canals**

The Dangerous Truth About Root Canals

Danica Collins |

What Are The Dangers of A Root Canal?

You may have a colony of bacteria in your mouth that you aren't even aware of, especially if you haven't heard about the dangers of root canals to your total body health.

Dr. George Meinig, a dentist and one of the founders of the American Association of Endodontists, asserts in his book *The Root Canal Cover Up* that a root canal is paramount to leaving an organ that dies to rot inside your body.

More than 40,000 root canals are performed in the United States every day.

When you get a root canal, it is usually because a tooth has decayed to the point that it is causing you severe pain. After the procedure, your tooth no longer hurts so you believe all is well. *Nothing could be further from the truth.*

What is a Root Canal?

During a root canal, the nerve and pulp are removed and the inside of the tooth is cleaned and sealed. Sealing this main canal does not account for the thousands of side canals that branch off each tooth.

These microscopic channels are left untouched and made up of dead rotting nerves. Bacteria feed off this dead tissue and the toxic waste created leaks into your bloodstream.

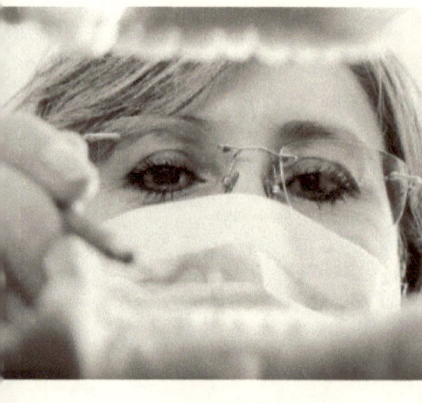

No matter what procedure or dental materials are used to perform your root canal, the resulting bacteria will leak into your system – how your body reacts to this threat is based on your immune system. The problems begin with the tubules located on the surface of your tooth. They are much like the pores on your skin. Your tooth looks solid but it isn't.

Oral bacteria found in the tubules move to the interior of the tooth and survive without a blood supply, on very little food and with no oxygen. Over time, they mutate into disease-causing bacteria and release toxins.

Antibiotics should help…right? *Wrong*.

The bacteria live in a protected colony that continually oozes toxins *out* but nothing – including medications – get *in*. It is much like a one-way street.

The bacteria thrive and continue to leak poison through the dead tubules into your bloodstream – but there is no blood supply to push antibiotics back through that same system.

Focal Infection from Root Canals

When a colony of bacteria exist in one specific location in your body – such as your bowels, tonsils or teeth – it can and will wreak havoc on a totally unrelated part of your body such as your brain, heart or joints. This is called focal infection. The longer it continues, the greater the chance that it will slowly weaken your immune system.

One of the dangers of root canals is that patients believe they've already addressed – and corrected – the problem by having the procedure done. Unknowingly, they've simply installed a condominium for bacteria.

Dr. Meinig explains, "This is not the usual medical story of a prolonged search for the difficult-to-find causative agent of some devastating disease. Rather, it's the story of how [bacteria] become entrenched inside the structure of teeth and end up causing the largest number of diseases ever traced to a single source."

The Dangers of Root Canals

- Arthritis
- Heart disease

- Neurodegenerative disease
- Digestion disorders
- Mental disorders
- Cancer

Root canal therapy

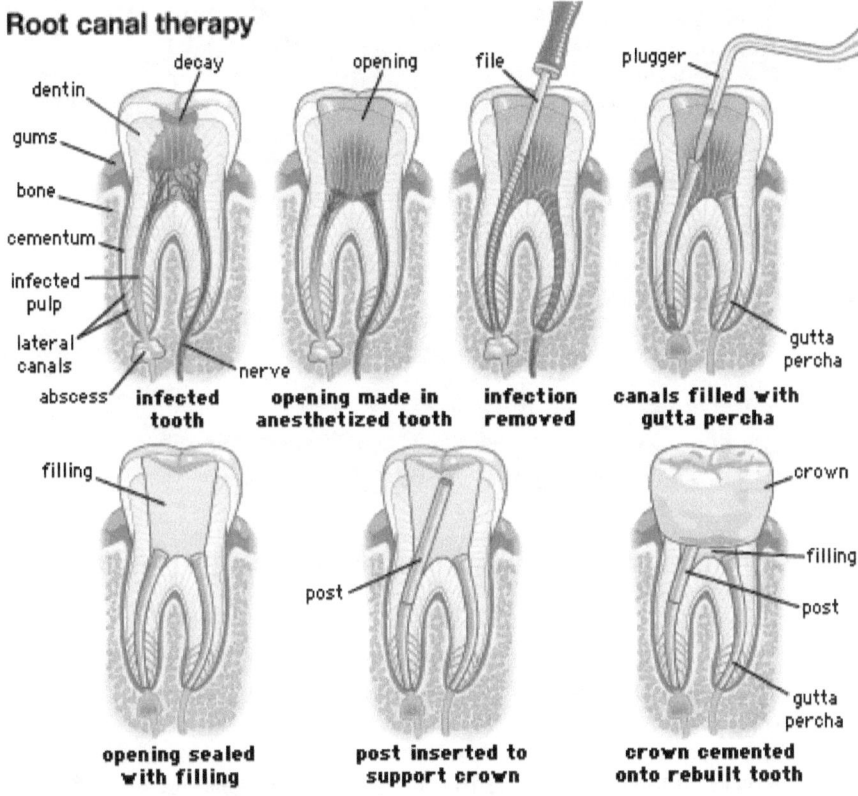

infected
tooth

opening made in
anesthetized tooth

infection
removed

canals filled with
gutta percha

opening sealed
with filling

post inserted to
support crown

crown cemented
onto rebuilt tooth

© 2008 Encyclopædia Britannica, Inc.

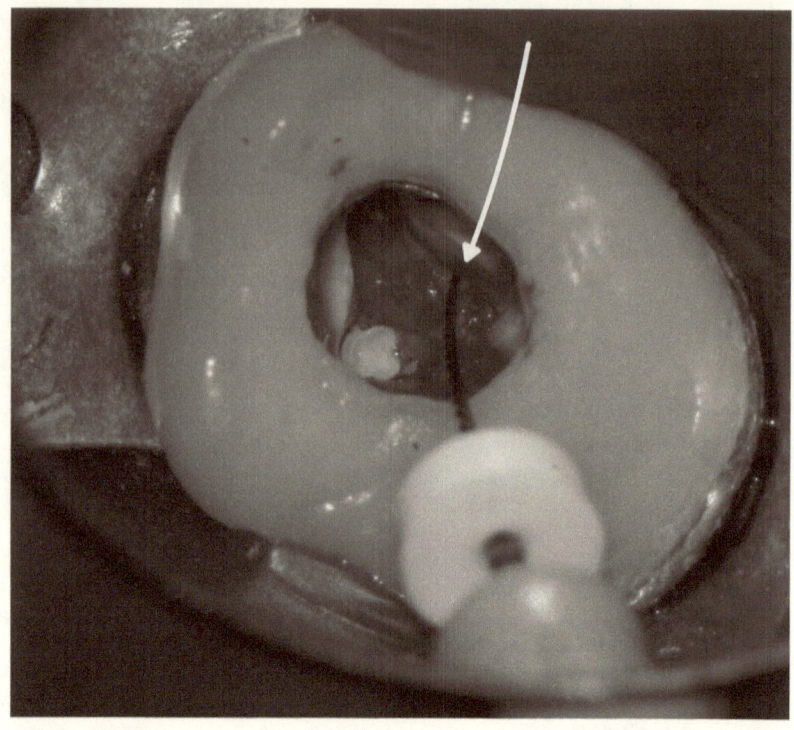

Bob Maddison has had two root canals done over the years, the first over 40 years ago.

Neither of those teeth remain in my mouth as both eventually broke apart. The worst one left just that metal post sticking up with the crown completely missing. That meant that the roots remaining in the gum had to be surgically removed. Also I am certain I recall that on one occasion a piece of tooth root was left in the jaw bone as it had broken off during the removal of that tooth.

They were quite nasty experiences as well as expensive and ineffective.

It is not possible for me to say what health issues resulted. However I have had numerous heart attacks, cardiac arrests, and open-heart surgery to repair a damaged mitral valve. Also 4 srokes have been diagnosed and treated in the last 4 years.

Does Root Canal Treatment Work?

By Dr. Shelley On January 5, 2012 ·
I often hear patients say, "My neighbor says to not get a root canal, because he's had three of them and each of those teeth have been pulled. Do root canals work?" Although root canal failure is a reality, it happens more often than it should. When a root canal failure is present, root canal retreatment can often solve the problem. This article discusses five reasons why root canals fail, and how seeking initial root canal treatment from an endodontist can reduce the risk of root canal failure.

The ultimate reason why root canals fail is bacteria. If our mouths were sterile there would be no decay or infection, and damaged teeth could, in ways, repair themselves. So although we can attribute nearly all root canal failure to the presence of bacteria, I will discuss five common reasons why root canals fail, and why at least four of them are mostly preventable.
Although initial root canal treatment should have a success rate between 85% and 97%, depending on the circumstance, about 30% of my work as an endodontist consists of re-doing a failing root canal that was done by someone else. Root canals often fail for the following five reasons:

1. Missed canals.
2. Incompletely treated canals – short treatment due to ledges, complex anatomy, lack of experience, or lack of attention to quality.
3. Remaining tissue.
4. Fracture.
5. Bacterial post-treatment leakage.

1. Missed Canals

The most common reason I see for root canal failure is untreated anatomy in the form of missed canals. Our general understanding of tooth anatomy should lead the practitioner to be able to find all the canals. For example, some teeth will have two canals 95% of the time, which means that if only one canal is found, then the practitioner better search diligently to find the second canal; not treating a canal in a case where it is present 95% of the time is purely unacceptable.

In other cases, the additional canal may only be present 75% of the time. The most common tooth that I find to have a root canal failure is the upper first molar, specifically the mesio-buccal root, which has two canals more than half the time. I generally find two canals in three out of four cases, yet nearly every time a patient presents with a failure in this tooth, it is because the original doctor missed the MB2 canal. Doing a root canal without a microscope greatly reduces the chances of treating the often difficult to find MB2 canal. Also, not having the right equipment makes finding this canal difficult. Not treating this canal often leads to persistent symptoms and latent (long-term) failure of the root canal. Using cone beam (CBCT) 3-dimensional radiographic imaging, like we have in our office, greatly assists in identifying the presence of this canal. In addition, when a patient presents for evaluation of a failing root canal, the CBCT is invaluable in helping us to definitively diagnose a missed canal.

The bottom line is that canals should not be missed because technology exists that allows us to identify and locate their presence. If a practitioner is performing endodontic (root canal) treatment, he or she needs to have the proper equipment to treat the full anatomy present in a tooth. Although getting a root canal from an endodontist may be slightly more expensive than getting one from a general dentist, there is a greater chance of savings in the long-term value of treating it right the first time.

2. Incompletely Treated Canal

The second most common reason that I see for root canal failure is incompletely treated canals. This usually comes in the form of "being short", meaning that if a canal is 23 millimeters long, the practitioner only treated 20 millimeters of it. Being short increases the chance of failure because it means that untreated or unfilled root canal space is present, ready for bacteria to colonize and cause infection.

Three reasons why a root canal treatment was shorter than it should be can be natural anatomy that does not allow it (sharp curves or calcifications), ledges (obstacles created by an inexperienced practitioner, a practitioner not using the proper equipment, or even an experienced practitioner in a complex situation), or pure laziness – not taking the time to get to the end of the canal.

Two factors that contribute to successfully treating a canal to length are proper equipment and experience. One example of proper equipment is an extra fine root canal file. Having the smallest most flexible root canal file (instrument used for cleaning) allows the practitioner to achieve the full length of the canal before damaging it in ways that are not repairable. If the doctor is using a file that is too large (and therefore too stiff) then he may create a ledge that is impossible to negotiate and will therefore result in not treating the full canal and could possibly lead to failure. Endodontists generally stock these smaller files, and general dentists often do not. Ledges can occur even with the most experienced doctor, but experience and the proper equipment will greatly reduce their occurrence.

The second factor that contributes to successfully treating a canal to length is experience. There is no substitute to having treated that particular situation many times before. Because endodontists do so many root canals, they develop a sensitive tactile ability to feel their way to the end of a canal. They also know how to skillfully open a canal in a way that will allow for the greatest success. Root canal treatment from an experienced endodontists greatly increases the chances that the full length of the canal will be treated and that failure will be reduced.

3. Tissue

The third reason I see for root canal failure is tissue that remained in the tooth at the time of the first root canal. This tissue acts as a nutrient source to bacteria that can re-infect the root canal system. Root canals naturally have irregular shapes that our uniformly round instruments do not easily clean. Two common reasons why tissue is left in a root canal is lack of proper lighting and magnification, which is achievable with a dental operating microscope, and a root canal that was done too quickly.
Immediately before filling a root canal space that I have cleaned, I stop to inspect the canals more closely by drying them and zooming in with the microscope to inspect the walls under high magnification and lighting. Even when I think I have done a thorough job, I will often find tissue that has been left along the walls. This tissue can be easily removed with experienced manipulation of the root canal file under high magnification.

The second reason why tissue may remain in a root canal treated tooth is that it was done too quickly. I am completely aware that the patient (and the doctor) want the root canal to go as quickly as possible, but one of the functions of the irrigant used to clean during root canal treatment is to digest tissue – the longer it sits there, the cleaner the tooth gets. This is good because areas that are not physically touched with a root canal instrument can still be cleaned by the cleaning solution. If a root canal is done too rapidly, the irrigant does not have time to work and the tooth does not become as clean as it possibly could be. Practitioners continually make judgment on when enough cleaning has occurred. Whereas we would love to have the patient's tooth soak for hours, doing so just is not practical. Therefore we determine when the maximum benefit has been achieved within a reasonable time period. If a root canal is done too rapidly and has not been thoroughly flushed then tissue may still remain and latent failure of the root canal may occur.

4. Fracture

Another common reason for root canal failure is **root fracture**. Although this may affect the root canal treated tooth, it may not be directly related to the root canal treatment. Cracks in the root allow bacteria to enter places they should not be. Fractures can occur in teeth that have never had a filling, indicating that many of them simply are not preventable.

Fractures may also occur due to root canal treatment that was overly aggressive at removing tooth structure. This is more common with root canals performed without magnification (such as the dental operating microscope) because the practitioner needs to remove more tooth structure to allow more light to be present.
Sometimes a fracture was present at the initial root canal treatment. When a fracture is identified, many factors go into determining if root canal treatment should be attempted. The prognosis in the presence of a fracture will always be decreased, but what we can never know is by how much. Sometimes the treatment lasts a long time, and sometimes it may only last six months. Our hope is that if root canal treatment was chosen to treat the tooth, then it will last a long time.

Fractures generally cannot be seen on an x-ray (radiograph). However, fractures cause a certain pattern of infection that can be seen on the radiograph which allows us to identify their presence. The cone beam (CBCT) 3-dimentional imaging system in our office can show us greater radiographic detail that helps us determine if a crack is present better than traditional dental radiographs. I have had many cases where I decided that root canal treatment or re-treatment would not solve the problem because the likelihood of a fracture was too high to justify treatment to save the tooth.

5. Leakage

The goals of root canal treatment is to remove tissue, kill bacteria, and seal the system to prevent re-entrance of bacteria. All dental materials allow leakage of bacteria; our goal is to limit the extent of leakage. At some unknown point the balance tips and infection can occur. The more measures we take to prevent leakage, the more likely success will occur. Four measures that can help reduce root canal failure due to leakage are rubber dam isolation, immediate permanent fillings, orifice barriers, and good communication with your general dentist.

Rubber Dam
A root canal should never be done without using the latex (or non-latex) barrier called a rubber dam. I was taught in school that root canal treatment without a rubber dam constitutes malpractice, and most practitioners would agree on that point. The rubber dam protects the patient in two ways. The first way that the rubber dam protects the patient is that it prevents small instruments from falling to the back of the mouth and being aspirated. The second way the rubber dam protects the patient is that it prevents bacteria rich saliva from entering the tooth and allowing for infection. A root canal done without a rubber dam is doomed to failure from bacteria. Although not required, use of the rubber dam at the time the access is restored can also hedge against failure from bacterial leakage. The first step to a successful root canal is to prevent the entrance of bacteria by using a rubber dam.

Permanent Filling (Build-Up)
When a root canal is finished by a specialist, it is a highly common practice for the endodontist to place a cotton pellet and a temporary material, which will then be replaced by the patient's general (restorative) dentist. This temporary material can begin leaking right away, but is generally sufficient for a period of 7-21

days while the patient makes an appointment with their general dentist.

The best way to reduce the chance of bacterial leakage is to have a permanent filling placed at the time root canal treatment is finished. This will assure that the tooth is sealed as much as possible against bacterial leakage. This filling is called an access restoration or a build-up. Although many endodontists place restorations to seal the access, many still place a temporary. Whether the patient receives a permanent filling or a temporary filling is largely dependent on a combination of factors including the practice philosophy of the endodontist, the preferences of the referring dentist, the complexity of the treatment plan, and the time allotted for treatment.

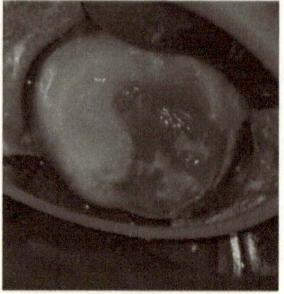

Orifice Barriers

When a permanent filling cannot be placed at the time a root canal is completed, an orifice barrier is the next best alternative. The opening to the canals is called an orifice, and the barrier can be a variety of materials. The material used in our office is a purple flowable composite that is bonded to the floor of the tooth and hardened with a high intensity light. Research will never prove whether this technique is effective or not in improving the long-term prognosis, but the general feeling in the endodontic community is that a bonded orifice barrier is better than nothing.

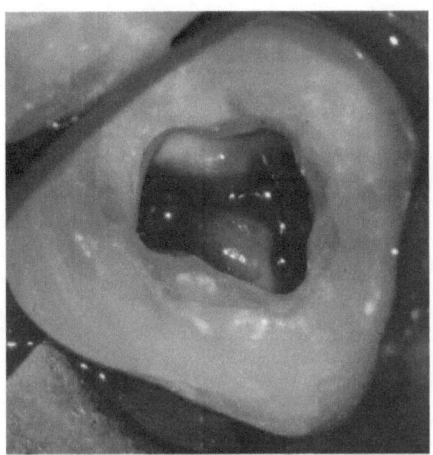

Good Communication and Timely Follow-up with the Restorative Dentist

Finally, leakage can be reduced when the patient sees their restorative dentist as soon as possible after root canal treatment has been completed. This can be accomplished when there is efficient communication between the endodontist and the restorative dentist. In our office we also send a monthly summary of patients to each doctor that they can use as one more layer to confirm that treatment on their patient has been completed and that the patient needs to be seen as soon as possible for restorative treatment. Much of the responsibility for timely restorative care is in the hands of the patient. Patients who delay restorative treatment after root canal therapy are risking failure of their root canal treatment, which may necessitate re-treatment at their expense. Patients should not delay in getting their root canal treated tooth permanently restored with a filling and in many cases with a crown.

The best way a patient can prevent failure of a root canal is to seek care from a practitioner like an endodontist that has experience, that has the proper equipment (including a microscope and possibly a cone beam CBCT 3D imaging), and to receive timely restorative treatment either at the time root canal treatment is completed or shortly thereafter.

To our readers: This article is number one in the world for people searching the topic 'Do root canals work?'

We Asked Dentists: Do You Use Mercury Fillings?

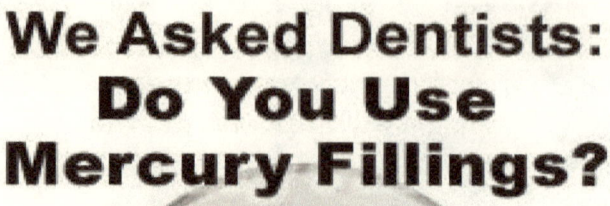

48% YES

52% NO

from Natural Life Magazine, January/February 1997
Mercury Fillings: A Time Bomb In Your Head
by Charles W. Moore

Sweden has banned mercury amalgam dental fillings, effective January, 1997, after determining that at least 250,000 Swedes have immune and other health disorders directly related to the mercury in their teeth. Denmark will ban amalgams beginning in January 1999.

In 1991, Germany's Health Ministry recommended to the German Dental Association that no further amalgam fillings be placed in children, pregnant women, or people with kidney disease, and in 1993 this was extended to include all women of child-bearing age, pregnant or not. Austria is also phasing out mercury fillings.

By contrast, the American Dental Association (ADA) says replacing amalgam fillings from non-allergic patients for the purpose of removing toxic substances from the body is "improper and unethical." The Canadian Dental Association (CDA) insists that there is no scientific evidence linking medical illness symptoms to mercury fillings, except relatively rare allergic sensitivity to mercury. (The number of persons with a specific and detectable sensitivity to mercury may not be so small. According to a Health Canada report, as many as 15 percent of people with amalgam fillings show signs of sensitivity to mercury. Some American researchers claim that at least 20 percent of people with amalgam fillings are "mercury toxic.")

What gives? Are the Europeans and Scandinavians hysterical Cassandras, in a sweat about nothing, or are the North American dental associations concerned about things other than patient health? Are mercury amalgam tooth fillings dangerous or not?

Amalgam tooth fillings are an alloy of 50 percent mercury, 35 percent silver, 13 percent tin, 2 percent copper, and a bit of zinc. Mercury toxicity was known in the 19th century, but amalgam's cheapness, ease of placement, and durability kept it popular. Dentists argue that mercury fillings last longer than resin composites, and are more gentle to tooth pulp. Composites also require more skill and time to place.

Mercury is a poison that penetrates all living cells of the human body. It is more toxic than lead, cadmium and arsenic. The smallest amount of mercury that won't damage human cells is unknown.

Unfortunately, mercury is a poison that penetrates all living cells of the human body. It is more toxic than lead, cadmium and arsenic. The smallest amount of mercury that won't damage human cells is unknown. Autopsy studies show a correlation between the number of mercury fillings and mercury levels in the brain and kidneys. Research also indicates that amalgams have an adverse effect on the immune system's T-lymphocyte count.

Scrap dental amalgam is classified hazardous waste by the American Environmental Protection Agency, and by law must be stored in unbreakable, sealed containers, and handled without touching. Dr. Sandra Denton, M.D., who specializes in treating chronic mercury toxicity, asks: "What is it about the mouth that makes this same stuff non-toxic?" Referring to American Dental

Association (ADA) claims that amalgams have been proved safe in studies, Dr. Denton challenges them to produce such studies. They have not. "On the other hand," says Denton, "research documenting mercury toxicity is voluminous." She has collected some 3,000 articles and several books on the topic.

A Danish study found that Multiple Sclerosis (MS) patients had eight times higher levels of mercury in their cerebrospinal fluid than healthy controls. An article in the *Journal of Forensic Medicine & Pathology* states: "Slow retrograde seepage of mercury from root canal or Class V amalgam fillings...may lead to multiple sclerosis in middle age." Dr. Hal Huggins of Colorado Springs, Colorado, a dentist who has MS himself, treats MS victims and people with other chronic health problems by removing mercury amalgam fillings as well as with detoxification and nutritional supplementation. He claims that 80 to 85 percent of his patients improve significantly.

Despite Huggin's successes, the U.S. Multiple Sclerosis Society opposes mercury amalgam removal, stating that they have found no scientific correlation between amalgams and MS. Dr. Huggins counters that if his results are to be written off as "anecdotal" or "placebo effect", then he has the largest collection of sustained recurring anecdotal placebo responses in the world.

Antibiotic resistant bacterial disease has become a significant and growing public health problem over the past decade. Studies show that genes protecting bacteria against mercury poisoning often bundle together with other genes that give bacteria antibiotic resistant qualities. If amalgam fillings stimulate and maintain populations of mercury-resistant bacteria, it's no major stretch to suggest that they might also be an agent in developing antibiotic-resistant bacteria. Research by Dr. Anne O. Summers, et al., at the University of Georgia shows such a relationship in monkeys. Dr. Summers put mercury fillings into the molars of monkeys. Within five weeks bacteria in the animals' intestines became resistant not only to mercury, but also to common antibiotics like penicillin, streptomycin, and tetracycline.

Another monkey study by Dr. Stuart B. Levy at Tufts University found that before having mercury fillings, an average of one percent of the monkeys' oral, and nine percent of their intestinal *Enterobacteriacae* were antibiotic-resistant. After receiving mercury fillings, 13 percent of oral and up to 70 percent of intestinal bugs became antibiotic resistant. The ADA responds by reiterating its stand that mercury fillings are safe, and arguing that animal studies "cannot be viewed as affecting humans."

It is well-established that elemental mercury vapour emits from amalgam tooth fillings during chewing, brushing, and eating hot and/or acidic foods.

It is well-established that elemental mercury vapour emits from amalgam tooth fillings during chewing, brushing, and eating hot and/or acidic foods. Most of this vapour is inhaled. allowing efficient absorption across the alveolar membrane in the lungs. Mercury easily crosses the blood/brain barrier – the brain and nervous system's main natural defense against many toxic substances. It can bind strongly to sulfur-containing proteins in nerve tissue (which may explain the association with MS – a disease of the nerve sheaths), and deposits in virtually all body tissues and organs. In experiments on mercury fillings in sheep, Dr. Murray Vimy, a dentist at the University of Calgary, proved that mercury migrates from the teeth into nearly all body tissues, especially the brain, kidneys, and liver.

The average dentist handles two or three pounds of mercury annually. According to *Consumer Reports*, up to 10 percent of dental offices have mercury vapour levels exceeding 50 micrograms per cubic metre of air – the upper limit considered safe for eight-hour workplace exposures. Dr. Sandra Denton cites a study at the University of North Texas that found neuropsychological dysfunction in 90 percent of dentists tested. Female dental personnel have a higher spontaneous abortion rate, higher incidence of premature labour, and elevated perinatal mortality, which has been substantiated by the EPA to be characteristic of women chronically exposed to mercury vapour. Stillbirths are significantly correlated with maternal blood mercury levels. Methyl mercury, the organic form of mercury that forms after oral ingestion of mercury, is 100 times more toxic than elemental mercury. Methyl mercury easily crosses the placental barrier and builds up 30 percent higher red blood cell levels in the unborn child than the mother.

The CDA counters that with billions of mercury amalgam fillings placed, there is no apparent epidemic of ill health effects. However, others argue that so many people have mercury fillings that no effective control group exists. Former Health Canada biologist Mark Richardson, who researched the scientific literature on mercury toxicity in preparing a risk assessment, notes that it is people wanting to maintain the status quo who conclude that there is no evidence that mercury toxicity is a health problem. He refers to the tobacco industry's stalwart insistence that studies linking smoking to lung cancer are unscientific. Richardson's report, under consideration by Health Canada, recommends limiting the number of mercury fillings per person.

Stubborn reluctance of dental associations to acknowledge the health risk of mercury toxicity from amalgam fillings may indeed have much in common with tobacco company tactics. If **diseases like Multiple Sclerosis, Chronic Fatigue Syndrome, and Multiple Chemical Sensitivity are linked to mercury exposure from tooth fillings,** significant potential exists for individual or class action lawsuits against dentists. Indeed, the German Dental Association has stated that if the government recommends further limitations on amalgam use, it will advise its members to stop using amalgams completely due to increasing risk of legal liability. The truth will eventually out, and if mercury fillings are indeed eventually proved harmful, a history of foot-dragging will not bolster the dental community's case in court.

Dr. Murray Vimy is certain that every time you chew, brush, or grind your teeth you absorb mercury. However, he councils against panic and suggests that mercury fillings be replaced with non-mercury materials like resin composites, porcelain, or gold, as needed. There is some risk that mass replacements could expose the patient to more mercury than if old fillings were left alone.

Charles Moore is a freelance writer living in rural Nova Scotia who specializes in health issues.

This is one of a limited number of articles from Natural Life Magazine presented on this website for free. To read more, please subscribe.

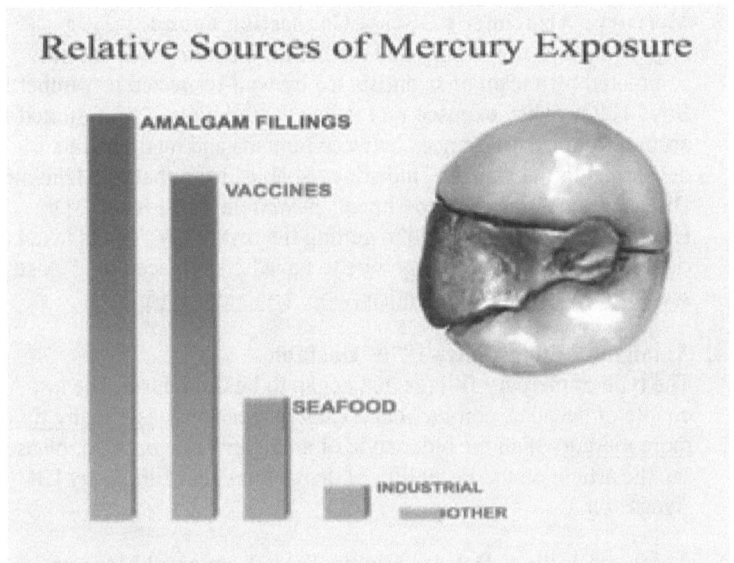

Relative Sources of Mercury Exposure

AMALGAM FILLINGS

VACCINES

SEAFOOD

INDUSTRIAL
OTHER

Amalgam / Mercury Dental Filling Toxicity

An often overlooked, but extremely important source of toxic material is the mercury from silver [mercury] amalgam fillings. Some people who are aware of the situation are confused by the mixture of information available. Unfortunately, statements from dental trade organizations and on a few poorly-researched news reports have muddled the situation.

Here are a few facts about mercury amalgam fillings:

1. **Causes Damage to Brain in Children**
   In February, 1998, a group of the world's top mercury researchers announced that mercury from amalgam fillings can permanently damage the brain, kidneys, and immune system of children.

2. **Amalgam Fillings Linked to Neurological Problems, Gastrointestinal Problems**
   The first large-scale epidemiological study of mercury and adverse reactions was recently completed and showed that of the symptoms looked at, there was a link seen to gastrointestinal problems, sleep disturbances, concentration problems, memory disturbances, lack of initiative, restlessness, bleeding gums and other mouth disorders.

3. **Mercury / Alzheimer's Disease Connection Found**
A study related to mercury and Alzheimer's Disease was recently completed by a team of scientists led by well-respected researcher Dr. Boyd Haley. They exposed rats to levels of mercury vapor diluted to account for size differences between humans and rats. The rats developed tissue damage "indistinguishable" from that of Alzheimer's Disease. Repeating the experiment showed the same results. Dr. Haley is quoted as saying "I'm getting the rest of my fillings taken out right now, and I've asked my wife to have hers replaced too." Also see: http://www.holistic-dentistry.com/artalzeimer.asp

4. **Amalgam Fillings Since 1970s Unstable**
The type of mercury fillings that began to be used during the last couple of decades, non-gamma-2 (high copper), releases many times more mercury than the older style of amalgam fillings. Also, please see the article on the instability of dental amalgam fillings by Ulf Bengsston.

5. **Amalgam Fillings Release Highly Toxic Elemental Mercury**
Mercury is one of the most toxic substances known. The mercury release from fillings is absorbed primarily as highly toxic elemental mercury vapor.

6. **Amalgam Fillings Largest Source of Mercury By Far**
Based on a number of studies in Sweden, the World Health Organization review of inorganic mercury in 1991 determined that mercury absorption is estimated to be approximately four times higher from amalgam fillings than from fish consumption. Recent studies have confirmed this estimate. The amount absorbed can vary considerably from person to person.

7. **Gold Crowns, Gum, Bruxism, Computer Monitors Increase Release of Mercury Significantly**
Gum chewing, grinding of teeth/bruxism, computer terminal exposure, presence of gold fillings or gold crowns (even if covering mercury fillings), teeth brushing, braces, and chewing cause the release of significantly increased amounts of mercury from the fillings. Also, please see the following short review related to increases in mercury exposure from dissimilar metals in the mouth, exposure to magnetic fields, chewing, etc.

8. **Cumulative Poison and Builds Up in Organs**
Mercury released from fillings builds up in the brain, pituitary, adrenals, and other parts of the body.

9. **Mercury Amalgam Fillings Effect Porphyrins**
   Preliminary results from the first <u>detailed biochemical analysis (scroll half-way down)</u> of patients who removed mercury amalgam fillings showed a significant drop in the excretion of porphyrins (important to heme synthesis -- heme carries oxygen to red blood cells), as well as a number of other key biochemical changes. Also, see the <u>Video</u> of the preliminary results from the study.

10. **Potential Contribuatory Factor in Other Diseases**
    Mercury from amalgam fillings has been implicated as a possible <u>contribuatory factor</u> in some cases of Multiple Sclerosis, Parkinson's Disease, IBS, reproductive disorders, allergies, and a variety of other illnesses.

11. **Mercury Build Up in Brain, Organs and Breast Milk of Fetuses of Mothers With Amalgam Fillings**
    Mercury from fillings in pregnant women has been shown to cause <u>mercury accumulation</u> in brain, kidneys and liver of human fetuses (all of the areas tested). Studies have shown that mercury can be passed to infants from <u>breast milk</u>.

12. **Proper Removal of Fillings Produces Eventual Health Improvement**
    A <u>recent study</u> published in the Journal of Orthomolecular Medicine related to the proper removal of mercury amalgam fillings from 118 subjects showed an elimination or reduction or 80% of the classic mercury poisoning symptoms. In many cases, it took 6 to 12 months after mercury amalgam removal for the symptoms to disappear.

13. **World-reknowned Experts Agree About Potential Danger**
    In contrast to statements from dental trade organizations, <u>toxicologists and medical researchers</u> are often quite concerned about the use of mercury. Lars Friberg, the lead toxicologist on the World Health Organization team looking at inorganic mercury and health effects recently stated that he believes that mercury is unsuitable for dental materials because of safety concerns.

14. **Canadian Class Action Lawsuit**
    Canadians are in the process of beginning a major <u>class action lawsuit</u> based on the fact that the government knew of but did not warn the public about mercury dangers from amalgam fillings. Legal actions related to mercury exposure from mercury amalgam fillings and vaccines are beginning in the United States. For more information and a directory of Mercury-free dentists, please see the <u>TalkInternation.com</u> web site.

Obviously, not everyone experiences acute toxicity effects from the mercury in amalgam fillings. However, virtually everyone does have mercury build up in their bodies from implantation of such fillings. The large increase in mercury exposure from the newer non-gamma-2 mercury fillings means that only time will tell how much damage has been caused by daily exposure to mercury to such fillings.

I do not recommend that people assume automatically that they will be healed by the removal of amalgam fillings. Many people are helped tremendously and some are healed. The 80% figure for people showing improvement within a year likely refers to people who had good reason to suspect that they were being significantly effected by the fillings. The percentage of people in the general population who might experience health improvement within one year after removal is probably much lower than 80%. I recommend going into the mercury amalgam removal procedure knowing that, at the very least, you will have removed yourself from a regular exposure to an extremely toxic material such that it will not build up in your organs and possibly cause significant health problems at a later date.

Mercury amalgam fillings should be removed **only** by dentists with experience using the IOAMT mercury amalgam removal protocol (presented with the permission of the excellent Preventive Dental Association web page). Such dentists are often experienced with proper evaluation and placement of composite fillings, both of which can be crucial for the success of the treatment. Biocompatability tests are often important in determining which composite materials can be safely used. I believe that composite (plastic) fillings are a better replacement than metal (e.g., gold) fillings even in chemically-sensitive individuals. They are, however, not without safety questions, but are still likely to be much less toxic than mercury amalgam fillings. Proper placement of composites should be left to experienced amalgam removal dentists as the average well-meaning dentist may not be aware of the newer placement techniques.

Further scientific information can be found at Mercury Adverse Effects Web Page, 150 Year's of Russian Roulette Web Page, Alt Corp's Amalgam Page, and Bo Walhjalt's Mercury Articles Web Page. More information about removal, detoxification, and placement of composite fillings can be found at Bioprobe, Inc. and at the Preventive Dental Association. Information about finding a dentist practicing non-toxic dentistry can be found on the Resources For Related to Non-Toxic Dentistry web page. Also, the AMALGAM mailing list can be a good source of accurate, up-to-date information.

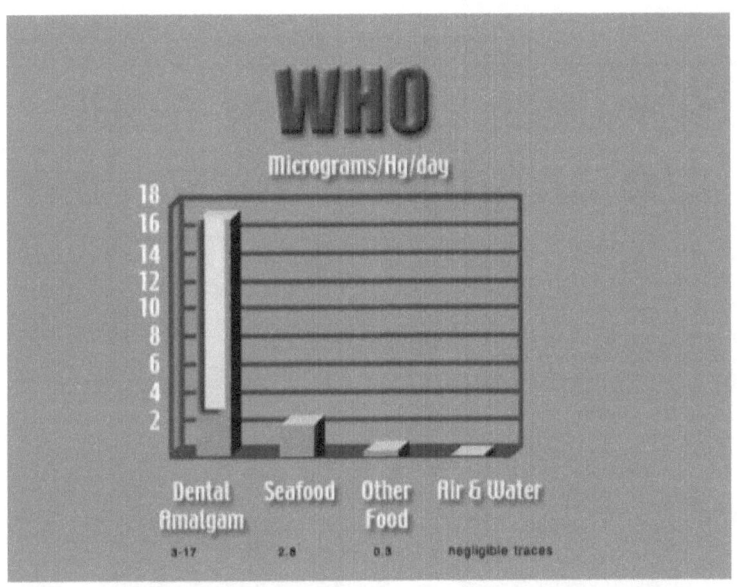

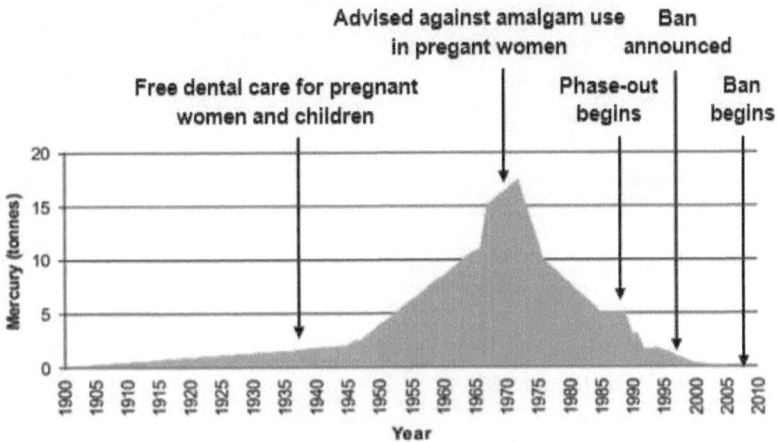

Mercury used annually in dentistry in Sweden

Saturday, March 22, 2008 | **East & Bays Courier**

Survivor gets her teeth into amalgam filling row

By Karen Kotze

The battle for or against amalgam fillings is still raging.

Juliet Pratt is one voice.

The single mum juggles sons, business and an active sports life.

She was also slowed by migraines, then swollen glands and glandular fever. She developed multiple chemical sensitivities, short-term memory loss, numbness, an irregular heartbeat, and a chronic cold that never cleared.

Mrs Pratt was eventually also diagnosed with chronic fatigue syndrome.

"I couldn't walk for more than 10 minutes at a time, and no amount of sleep relieved the tiredness," Mrs Pratt says. "It was utterly debilitating."

A solution took years.

"Years of misdiagnosis, of being told I was imagining things, of tests and being told there was actually nothing wrong with me, other than 'getting older'," she says.

Mrs Pratt believes many of her problems were caused by amalgam fillings.

The fillings contain mercury, which is toxic, Mrs Pratt says, and can be absorbed into the bloodstream.

She claims more than half of American dentists are now mercury-free and that Sweden placed a total ban on all amalgam fillings in 1993.

"The Swedish government pays 70 percent of the costs of removing the fillings. Also, in 1993 Germany's largest manufacturer of amalgam stopped making it," she says.

"The argument that it is fine because it has been used for 150 years makes

Amalgam nightmare: Juliet Pratt shares her story of recovery.
Photo: KAREN KOTZE

no scientific sense. We have abandoned other remnants of pre-civil war medicine and all other uses of mercury, but not dentistry?"

At one stage Mrs Pratt had 40 amalgam fillings in her mouth but she was told the recommended maximum was eight.

She says people should research links between mercury and alzheimers, arthritis, asthma, cardiovascular disease, depression, autism, multiple sclerosis, liver disorders and parkinsons.

"Significant numbers of alzheimers patients have shown abnormally high levels of mercury in areas concerned with memory," Mrs Pratt says.

"The economic impact that these diseases will have on New Zealand will be dramatic in years to come," she says.

Mrs Pratt is a financial adviser and a member of the Million Dollar Round Table. The organisation, founded in 1927, sets the standard for excellence in the financial profession.

"I am focused on finding sponsors for the Lake Taupo Cycle Challenge in November. I intend to challenge four celebrities to relay the 160km course. I want to set up a register of people in New Zealand who have been affected by mercury amalgam and present it to Parliament. My aim is to ban the fillings in New Zealand," she says.

She asks anyone who has been affected by what they believe is mercury amalgam fillings to email her on juliet.pratt@xtra.co.nz, or call 359-9911 or 0275-309-319. For more information visit www.mercury free-now.com.

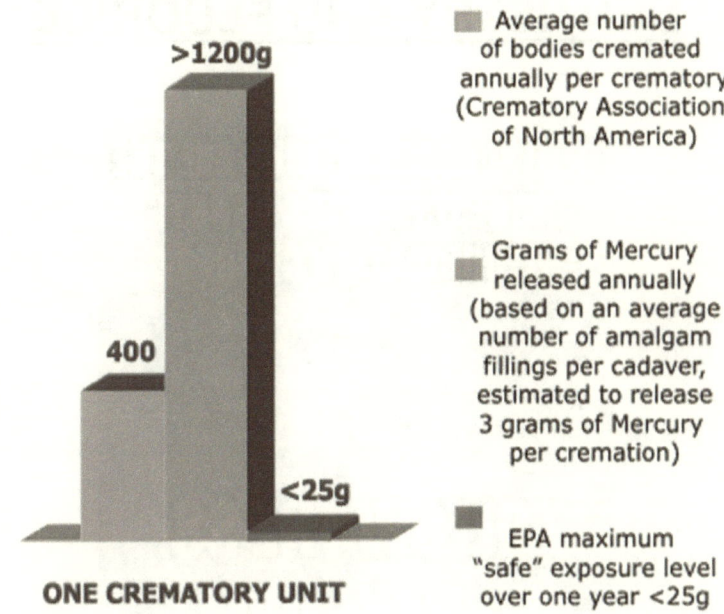

ONE CREMATORY UNIT

Average number of bodies cremated annually per crematory (Crematory Association of North America)

Grams of Mercury released annually (based on an average number of amalgam fillings per cadaver, estimated to release 3 grams of Mercury per cremation)

EPA maximum "safe" exposure level over one year <25g

Unfortunately yet another way mercury is released into the atmosphere to cause on-going health issues to the community.

Major pathways of mercury due to use of dental amalgam every year[viii]

Major release/pathways	Mercury (metric tonnes/year)
Atmosphere	50-70
Surface water	35-45
Groundwater	20-25
Soil	75-100
Recycling of dental amalgam	40-50
Sequestered, secure disposal	40-50
Total	260-340

Source: UNEP

DENTISTRY AND FLUORIDE

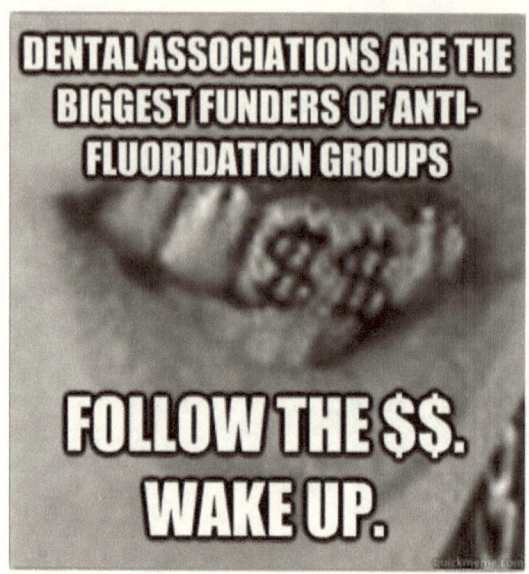

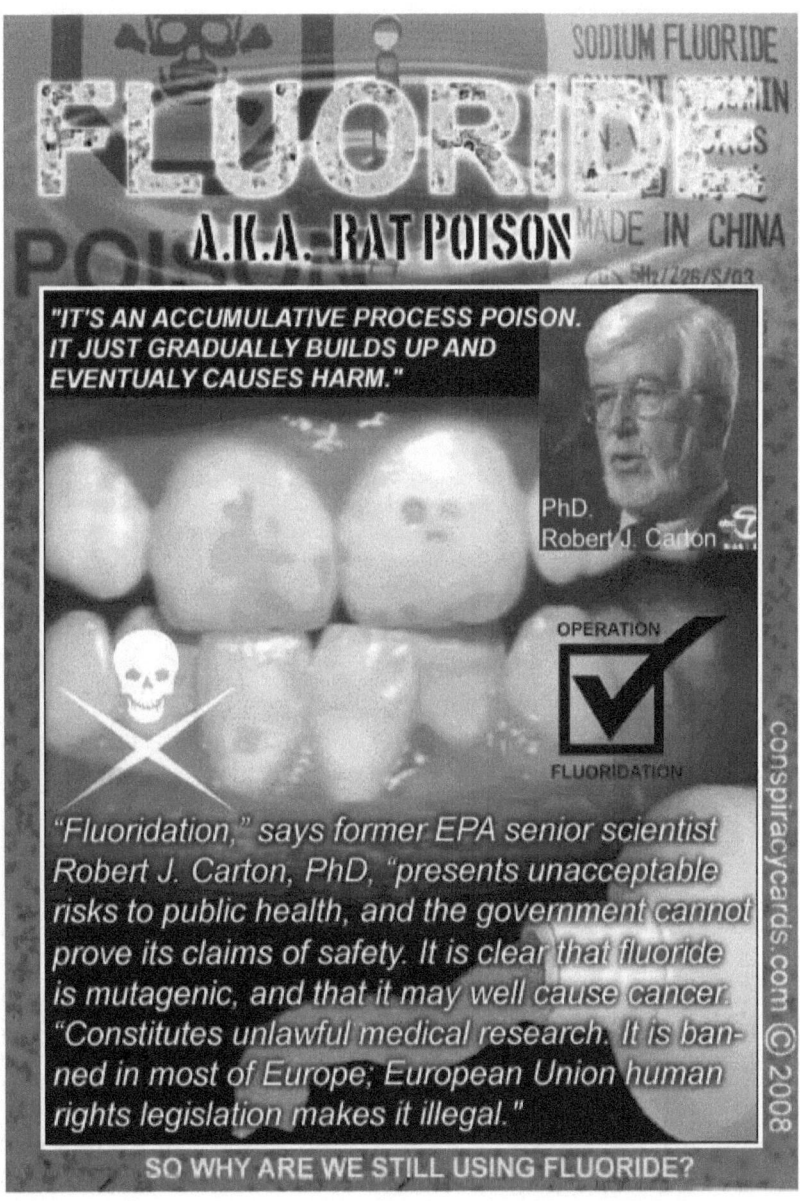

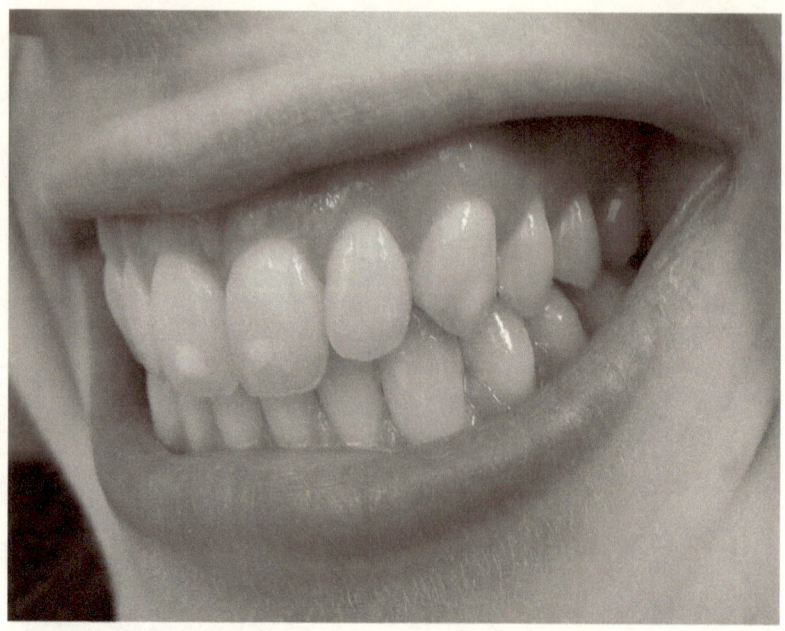

This picture shows the result of Fluoride on teeth. You need to see it in colour on Internet to fully appreciate.

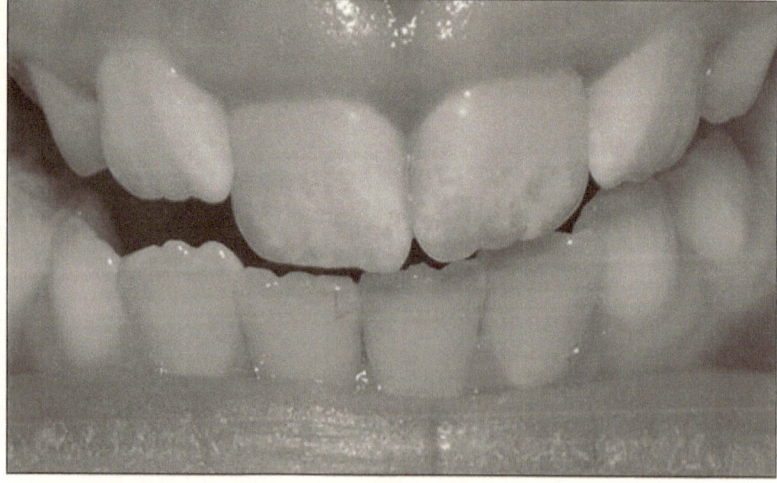

This is what the dental profession and associations are doing to our (children's) teeth. Worse is in store for human bones.

Bearing in mind the disasters caused by this "Dental Profession" for over 100 years, relative to fiasco of dental amalgam (mercury) use, root canal treatments, lack of integrity or care in such canal root technology, how or why should we listen to, or accept their endorsement of yet another proven deletrious chemical?

Yet these are they who loudly shout the efficacy of fluoride being imposed upon entire communities. Why is that? Seriously ask: "why is that?" Then ask "why do they actively promote mercury as a recommended treatment for dental fillings?' These are they who bad-mouth and decry those who have done the appropriate research to establish the danger and hazards of both fluoride and mercury that they actively use and promote. Frankly its not only disgusting, but criminal, given today's knowledge.

If "Fluoride" was the only irrevocable sin in which this profession was involved, it may be perhaps classed as "one short-coming" or error. But the information above indicates this is not the case.

It is either a part of a 'higher' (actual, not 'theory'. – see my books.) "conspiracy", or an intentional disregard for human health and safety.

The facts are simple and easily correctly established. Just do an Internet search to validate all evidence.

Frankly we have been betrayed, lied to, had all actual truth and evidence concealed and denied to us. **"The evidence is out there."** All you need to do is find it.

So, we have been lied to or mislead (intentionally) about 'root canals', amalgam and mercury. And this 'fluoride' issue.

CAN THIS DISASTER GET ANY WORSE?

Oh yes it can. The "unholy" disaster of the dental profession merged almost unseamed with the entire medical profession. Organised and government supported death and disaster was in fact <u>imposed</u> upon the civilian populations.

It would start with our water supply, our toothpaste, then our health services. But not necessarily in that order of implementation. Yes, "implementation", for others and I see a definite "plan" being implemented.

<u>This is NOT a conspiracy "theory", but an actual chain of events that is undeniable and real.</u>

<u>This is "their" attack/rebuttal etc. by Bob Maddison</u>

"they" denigrate "us" using terminology.

"Denigrate" (Chambers English Dictionary) v.t. to blacken (especially a reputation).

The words "**Theory**" or "Theorem", Chambers English Dictionary:

"a demonstrable or established but not self-evident principle: a proposition to be proved. – not practical: speculative – the speculative parts of a science, - speculation – to form opinions solely by theories, to speculate."

Frequently those who oppose conventional academia are called as referring too "pseudo science".

"Pseudo (Science)"` Chambers English Dictionary. "in comparison, sham, false, spurious: deceptively resembling; - pretentious –etc. These words are among the favourites of the abusers. However their pronouncements are in reality the "pseudo science" and totally false.

We can clearly see therefore, once the evidences as in the above writings is clearly understood, that the "pseudo science" is entirely held by those who assert that mercury amalgams, root canals, fluoride, and vaccines are safe and effective. Also we see that their stance is based entirely "theory" with absolutely no valid evidence, **and that constitutes the real "conspiracy theory".**

The "academics", those who have their credentials diplomas etc. love to refer to "us" as "unqualified". Etc. as though they are somehow superior. This is false as learning is not "education". Anyone can read and learn. But "they" choose to down-play the bulk of mankind. I get annoyed, but refer to my earlier books for evidence etc.

BOB'S COMMENTS (from "It Ain't Necessarily So")

Hereunder is the response of a typical "de-bunker", and as usual, the style is to attack the individuals, belittle, put down etc. Sad.

"My greatest qualifications are a keen eye, good sense and the power of reason. When the hoax advocates make claims that are based on flimsy evidence, sloppy research, and a poor understanding of the sciences, it does not take a PhD to figure out they are wrong. In general, the main proponents of the hoax theory are people who have no special education, training or experience to qualify them to make their claims. They are in no better position to judge the facts than you and I; so use your own sense of reason."

(Hey, as was once wisely said, I have no "qualification" but I can read. (And see.) The evidence used is all sourced from NASA official releases, so that is neither "flimsy" or "sloppy". To compare any two photographs does not   require any "special education" and **to infer "we have no "special education" is merely saying 'I have credentials, you have**

none. So shut up.' Oh dear, I forgot. We are supposed to unquestioningly accept and believe whatever "they" tell us.

From henceforth the time has come to refuse to accept or countenance the use of the words or name "conspiracy theory", but to accurately name and call such things "Actual conspiracy", while referring to those who attempt to denigrate the previously used words and people as "Conspiracy deniers or ignorants". Thus there will be no "conspiracy theorists", but from now only those "aware of actual conspiracy".

There is a blatant and obvious attempt by many to denigrate, slander, ridicule, and mock the reputation of people or groups that openly proclaim that there is a real and actual conspiracy relative to many issues. It is done loudly and publicly by those who presume they are the only acceptable authorities relative to the matter or subject being denigrated by them. Lets look at the situation of this.

It is a very old and well known technique used by those who assume they are "credetialled" so called "experts" and "authorities". I have this to say about them now, as I did when I wrote the following in 2009. ("I Can See Clearly Now" written by Bob Maddison P.159-))

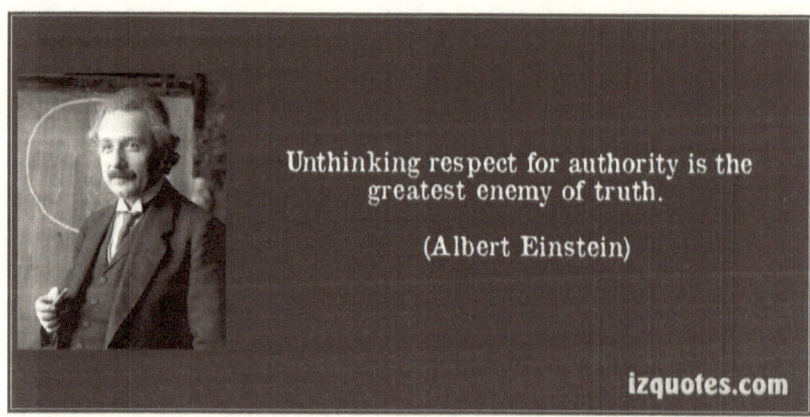

Unthinking respect for authority is the greatest enemy of truth.

(Albert Einstein)

izquotes.com

THE UNCHANGED HUMAN CONDITION

It also gives us a very special, secret pleasure to see how unaware the people around us are of what is really happening to them." ~Adolf Hitler

Is it that we humans as a species just do not learn from past obvious mistakes, or is it that we have been designed with a defect in our mental ability? It may be because a little of each element is involved that we find ourselves in our condition. In one section of my writings we look into the "who or what" is responsible for our very being and make up as well as the "purpose" for our being on this planet. If that is an accurate proposition then it would validate the claim that the human species has all the hallmarks of a domesticated species.

Regardless of the nature of our origin or purpose (if any) for being on this planet, it would appear to me that there is in place a cunning, carefully designed and implemented plan (or plot) to direct our path and mould or shape our collective races in a specific manner. Call it the mother of all conspiracy theories if you will. If all of this is totally wrong, it makes an interesting interpretation out of a lot of given facts and established data, and should be entertaining reading. When one looks at the evidence that will unfold in the next several sections one may not only agree with that premise, but one may also start to feel rather insecure and afraid. It is not the purpose of this book to frighten or alarm, but to open our eyes, make us aware of possibilities, forewarn us, and show that regardless of our worst nightmare becoming manifest, there is truly light at the end of the "tunnel".

For a long time now, in some cases for many decades, various voices have been raised in protest and to enable public awareness of various real dangers being foisted upon humanity. They have largely been ignored, and the small percentages that have not ignored the warnings but joined the aware group of protesters have been unable to cause any change to the adverse situations. Frankly we have been "attacked" on too many fronts simultaneously and the defending ranks are too thin to enable effective defence in so many diverse directions as to ensure safety for our species. Warnings about most obvious and in many cases self-evident dangers are blatantly ignored by the "powers that be", those who could make changes. Over the course of these writings we will identify the "powers that be".

The absolute dangers to mankind in the use of fluoride in water supplies, amalgam fillings (loads of toxic mercury) in dental work, vaccines, chemotherapy, radiotherapy, electroshock treatments, microwaves (not only in stoves), and a host of similarly dangerous issues have been not only ignored, but actively and willfully suppressed. (If you doubt this, then do some Internet research to establish all issues and concerns.) *These and other issues are loudly and vehemently declared as quite safe for use or implementation by the "authorities". The media carries that biased one-sided opinion, the dictates of "authority" without question and gives little to no coverage of the nature and issues of concern. This has been going on for decades, it has not changed and probably will never change. Protest is virtually useless. Why is this? It becomes apparent "things" are controlled.*

It is as though there really is an orchestrated assault against mankind on as many fronts as can be found.

"Authority" goes unquestioned in or by the media, allowing the most ludicrous of claims to become established as indisputable facts in the minds of the populations. It is well known that if someone shouts the most obvious of lies loud and long enough it will come to be accepted by most.

Therein is a key to a lot of problems. A lot of people simply do not think for themselves, and are accepting of what "authorities" dictate. Think about this. What is an expert or authority in most cases? Most are merely people who have gone through some educational institutions and learned what to accept, what to think, what to say, and having proven by examination that they can effectively recite the dogma of their branch of study, they are awarded some certification to that effect. These then become new peers to the closed group that monitor all aspirants and venturers into their field of "expertise".

Thus such a self-perpetuating closed group of "experts", custodians of the sacred dogma, forever control any given branch of learning or endeavour. Thus they can ignore protest with impunity, for they and their gospel will outlive any given protesters. A good example of this in action is the age of the Sphinx issue. The evidence of geologists presented decades ago is still totally ignored by "Egyptologists", who shifted the discussion to how "insensitive" the protesters were. The real issues have never been addressed or answered, as ignoring them has proven the most effective of manner in which to deal with such embarrassing (for them) questions or issues. This same tried and proven effective method is used with regular monotony whenever experts or authority (political leaders are skilled in its use) are faced with indisputable valid evidence of error in their thinking and dogma.

We see questions and error ignored in many fields. Theory of evolution is one dealt with in another book. "Big Bang" theory is quite nonsense and impossible to reasonably understand or accept. There are numerous cloudy issues we are asked to blindly accept because we are told to and quite simply that is the way things are. If "they" speak, we are expected to remain silent and accepting for the thinking has been done on our behalf. Religion is rife with "don't think about it" issues.

Why don't most people think?

*We will look into that very basic question in a section that deals with the **education** system imposed upon us by our diligent and vigilant government authority. We are carefully brought up and educated so as not to think independently. Our common **religions** train us in acceptance and not to think for ourselves. And should you join any **military** force, then in most cases and issues you become forbidden to think.*

The populations at large are mostly compliant and subservient to the authority and experts that they enable and allow to rule and preside over them. We will look into this issue as well. We will see that there may indeed be "something in the water" that should give us all cause for major concern. Yes collectively we drink the reticulated water, eat the processed foods, have our vaccine shots, take the medications of corporates, then wonder why there is an epidemic of early deaths or the populations are obese and unhealthy. We are addicted to and enslaved by an epidemic of alcohol, tobacco, and addictive drugs. Organized crime is almost in control in some areas.

*Most **religions** are far from user-friendly. We will look at this claim in more detail later and see that untold human misery and death result because of this insidious control method. Now understand that there is a huge difference between "religion" and "**spirituality**". I am not against the latter, but assuredly have no more time for the former. Essentially "spirituality" is what one has or feels from "within", it is an expression of self. "Religion" is from "without", extraneous to the self and is a set of proscribed scripture, doctrine, dogma, beliefs, creeds, etc. It is control and regulation. It is rarely conducive to peace and harmony. Also understand that I am not against "GOD", but that I have no time for and* *am against those named entities proclaimed as gods by the many "religions" or sects. "GOD", or the intelligent fabric of the universe has absolutely nothing whatsoever to do with religion. Yes I know we have all been brought up and trained quite differently than to even think that kind of thought.*

A huge troublesome issue with most religions it that they are frequently used as a vehicle by some to impose their will and way (often narrow, self-promoting, sometimes nefarious) upon others. Such will generally find within their acceptable scriptures some text or quote, interpret it (generally out of context) to validate or justify the imposition of their will to establish their ends. They loudly proclaim that it's from the very words of god, and thus must not be questioned but obeyed and implemented. Thus "holy" crusades were launched and entire cities and populations perished. ___Too many people let the religions and their leaders do the thinking for them, just as they do with so called health, medical, scientist, and a host of government and other "experts" and "authorities".___

*Additional to all the above controls over mankind we are also subject to others, and one of the more far-reaching and devastating is the use of almost total control of "**finance**" to dominate and control the bulk of humanity. ....*

So there we have a simplified overview of some of the many issues and problems assaulting mankind. Not only have we been seriously cheated, but also we have been misled and blatantly told great lies.

*Generally individuals do not even know their true nature, their actual and real identity. This has been withheld from them. They believe they **are** a mortal body and being. They are vaguely told they "have" or will have a spirit and even its fate is determined by their degree of obedience and subservience. They are governed property. They are compliant, obedient and blindly accepting of whatever is told or given them. They don't know what they want or what to do with this life, yet look forward to another life that will be "eternal". They are taught that they must pray to some god, that they are "sinners", need forgiveness always and need his mercy and grace. They have no clear understanding of what the whole "god" thing is all about, and thus know almost nothing of reality. They trust the religions and its leaders for their vague future salvation and pleasure in an ill-defined heaven, if they've paid their "dues". They view this "hologram" of a world as a reality and look for a god as "the man* *behind the curtain", one who will make it right and see to justice in the next world or life.*

Generally fearing death, they do not know how to die with grace or dignity, and often do not know how to live with grace and dignity. Worst of all, few will ever question the "status quo" and seek to understand anything that is not available at the mall."

(Written by Bob Maddison, probably October 2010, published in "It Ain't Necessarily So".)

Active ingredients
Sodium fluoride 0.24% (0.14% w/v fluoride ion)...
Triclosan 0.30% ...

Uses aids in the prevention of: • cavities

Warnings Keep out of the reach of children
swallowed, get medical help or contact a Pois
 • bleeding or redness lasting more than 2 wee
  These may be signs of periodontitis, a serious

Use helps protect against cavities

Warnings
Keep out of reach of children under 6 yrs. of age. If more than used for brushing is
accidentally swallowed, get medical help or contact a Poison Control Center right away.

Directions • adults and children 2 yrs. & older: brush teeth thoroughly after
meals or at least twice a day or use as directed by a dentist   • do not swallow
 • to minimize swallowing use a pea-sized amount in children under 6
 • supervise children's brushing until good habits are established ▼

9516161872

Yet, this is OK in 1-2 glasses of
Fluoridated Water?

I notice that now un-fluoridated toothpaste is no longer available on Super market shelves in the entire Wellington region. Also quite nefariously I see that most if not all brands currently available has had the warning notices on the back of the tubes removed. Of course they're getting away with this as no one seems to really care.

Obviously the fluoride is working fine for its real purpose.

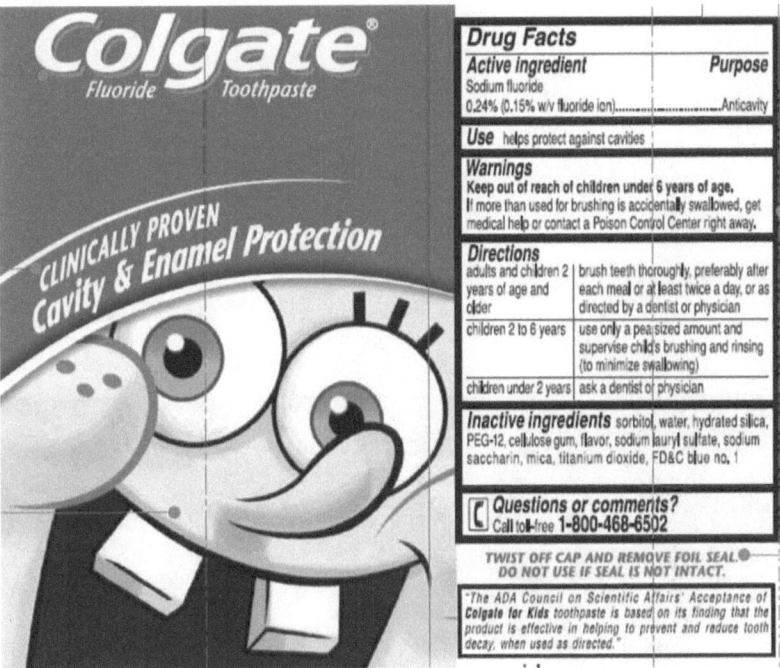

So why cannot people see this obviously available information and either boycott such products and services, or loudly protest? It turns out there is a large protest group and informed people. But they are silenced and not given coverage or press by the controlled media. These are the informed that are mocked as "conspiracy theorists" etc as mentioned above.

These are the "29%" who vote "no" to fluoride use in seemingly all reported results for referendums. Strangely consistent % figures everywhere – makes me suspicious of results.

The Myth

"harmless"

The reality

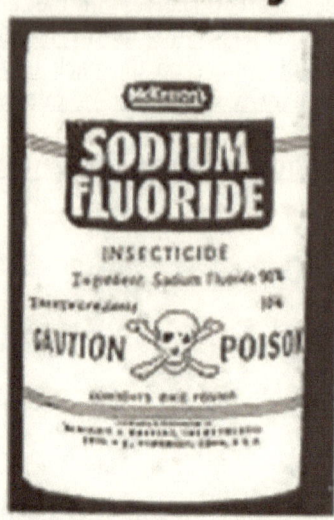

"think again"

Conspiracy Theorist:

Nothing more than a derogatory title
used to dismiss a critical thinker.

www.exposingthetruth.co

Fluoride pushers know they are pushing poison

I have met lots of doctors and dentists who advocate water fluoridation. Every one of them is fully aware that it isn't "natural" fluoride being pumped into the water. They know it's a harmful chemical, but they push it anyway. That makes them all **"science" criminals** who are knowingly harming people. Just like GMOs and vaccines, water fluoridation poisoning is all conducted in the name of "evidence-based science."

But all the evidence actually shows **hydrofluosilicic acid (flouride) is a deadly poison**.

Learnmore: http://www.naturalnews.com/039776_fluoride_hydrofluosilic ic_acid_insecticide.html#ixzz2ayd3p68Q

'Fluoride Free' Cornwall Will Improve Community Health

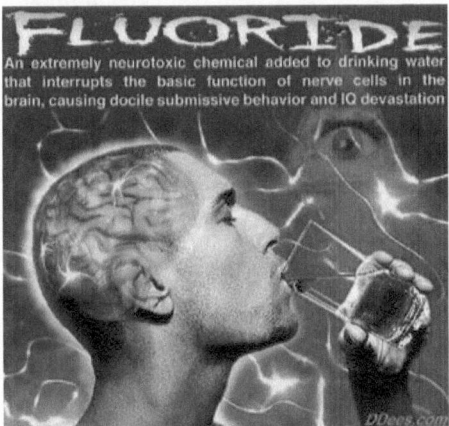

I was pleased to see that Paul Brisebois and his supporters made a presentation to City Hall regarding the perils and risks of fluoridated water.

The information about fluoride is getting out there. One only has to Google the Internet and you will find many, many valid reasons for avoiding fluoride in water, beverages and toothpaste.

Basically, fluoride is a slow killer, and causes a deterioration of health over the long term, and does nothing to improve the health of one's teeth. Fluoride, a bye product of the fertilizer and aluminum industries, has made a lot of wealthy families very rich, including the powerful Mellon family in the U.S. Why dispose of an industrial waste product when you can sell it to municipalities and toothpaste companies and make a bundle.

It is sobering to realize that the original "quacks" were dentists who advocated the use of mercury amalgam and that most dentists are still advocating it today. --- *The maximum amount of mercury that the Environment Protection Agency allows people to be exposed to is 5,000 times smaller than the permissible amount of lead exposure; in other words the EPA apparently considers mercury to be 5,000 times more toxic than lead."* **Marcia Basciano DDS**

"Primates that received just ONE vaccination containing thimerosal, the mercury-preservative found in many vaccines including the new swine flu shot, had significant neurological impairment when compared with those who received a saline solution injection or no injection at all." <u>Doctor Joseph Mercola MD</u>

"Mercury is a powerful poison. Published research has shown that mercury, even in small amounts, is more toxic than lead, cadmium and even arsenic. Some of the most common signs and symptoms of mercury exposure include irritability, fits of anger, lack of energy, fatigue, low self-esteem, drowsiness, decline of intellect, low self-control, nervousness, memory loss, depression, anxiety, shyness/timidity and insomnia." <u>**Eric Davis, DDS**</u>

"In spite of well-established health risks, organic mercurials are still added to prescription and non-prescription drugs, such as medicines for hemorrhoids (Preparation H), as well as in formulations for the treatment of bacterial and fungal infections. The ban on thimerosal in contact lens solutions did little to eliminate its use in other products, such as eardrops and nose drops. Thimerosal continues to be used today in a variety of health-related products: for preserving vaccines and intramuscular injections, cosmetics, and some drugs that must be kept in solution. It is the thimerosal used in childhood vaccines that gives the greatest cause for concern." <u>**Eric Davis, DDS**</u>

<u>**Note MOST "flu shots" contain 25 mcg of mercury in a SINGLE dose! ALL "flu shots" contain mercury.**</u>

There is **NO LEVEL** of mercury in your body that is safe. One microgram is about 4.3 Billion atoms of mercury each of which can harm you.

MERCURY DENTAL FILLINGS
By the Numbers

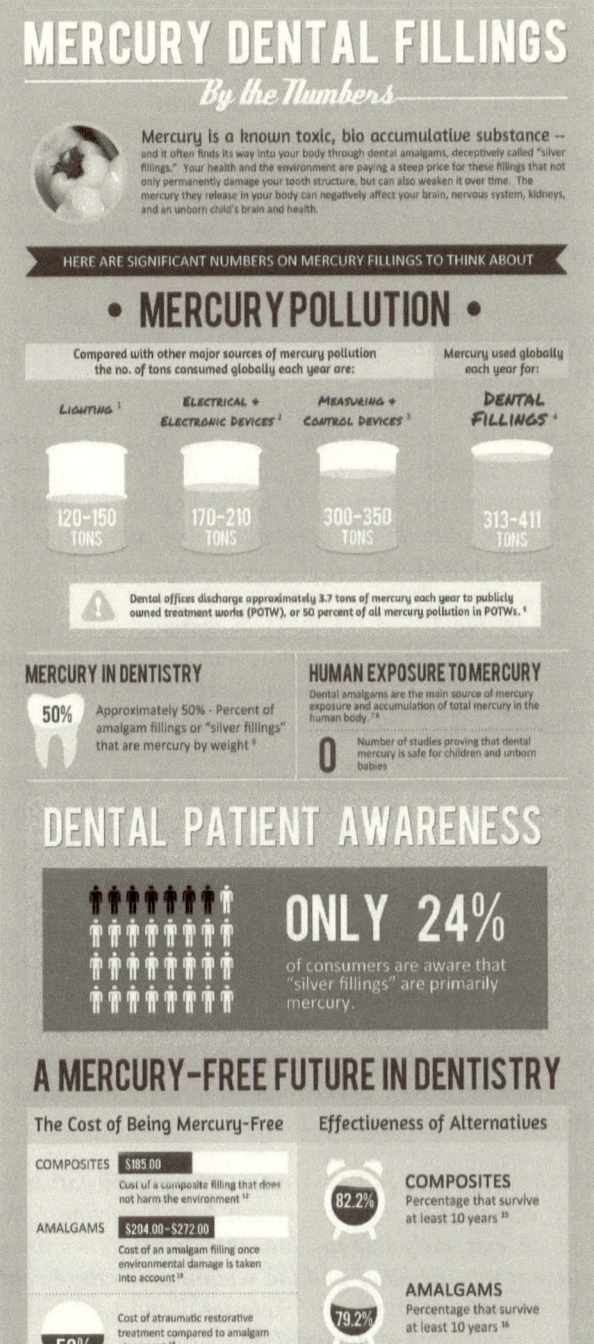

Mercury is a known toxic, bio accumulative substance -- and it often finds its way into your body through dental amalgams, deceptively called "silver fillings." Your health and the environment are paying a steep price for these fillings that not only permanently damage your tooth structure, but can also weaken it over time. The mercury they release in your body can negatively affect your brain, nervous system, kidneys, and an unborn child's brain and health.

HERE ARE SIGNIFICANT NUMBERS ON MERCURY FILLINGS TO THINK ABOUT

• MERCURY POLLUTION •

Compared with other major sources of mercury pollution the no. of tons consumed globally each year are:

Mercury used globally each year for:

LIGHTING [1]	ELECTRICAL + ELECTRONIC DEVICES [2]	MEASURING + CONTROL DEVICES [3]	DENTAL FILLINGS [4]
120-150 TONS	170-210 TONS	300-350 TONS	313-411 TONS

Dental offices discharge approximately 3.7 tons of mercury each year to publicly owned treatment works (POTW), or 50 percent of all mercury pollution in POTWs. [5]

MERCURY IN DENTISTRY

50% Approximately 50% - Percent of amalgam fillings or "silver fillings" that are mercury by weight [6]

HUMAN EXPOSURE TO MERCURY

Dental amalgams are the main source of mercury exposure and accumulation of total mercury in the human body. [7,8]

0 Number of studies proving that dental mercury is safe for children and unborn babies

DENTAL PATIENT AWARENESS

ONLY 24%
of consumers are aware that "silver fillings" are primarily mercury.

A MERCURY-FREE FUTURE IN DENTISTRY

The Cost of Being Mercury-Free

COMPOSITES	$185.00

Cost of a composite filling that does not harm the environment [12]

AMALGAMS	$204.00-$272.00

Cost of an amalgam filling once environmental damage is taken into account [13]

50% Cost of atraumatic restorative treatment compared to amalgam placement [14]

Effectiveness of Alternatives

82.2% **COMPOSITES** Percentage that survive at least 10 years [15]

79.2% **AMALGAMS** Percentage that survive at least 10 years [16]

"Mercury disconnects the intelligent organizing force that organizes your biochemistry." Doctor Dietrich Klinghardt

"The epidemiological evidence is compelling and statistically conclusive we found that the prevalence of speech disorders, autism and heart arrest was a function of the mercury dose that the children received." Geier and Geier

"Accumulation of mercury in the body is associated with accelerated atherosclerotic progression in men." The Kuopio Ischemic Heart Disease Risk Factor Study

"Mercury induced reactive oxygen species and lipid peroxidation (forming free radicals) has been found to be a major factor in mercury's neurotoxicity, along with its leading to decreased levels of the vital enzymes glutathione peroxidase and superoxide dismutase (SOD)." Willis S. Langford

"A study led by the U.S. EPA looks at mercury levels in women of childbearing age. According the study, one in ten women nationwide exceeded the mercury concentration levels that risk fetal health." Environmental Health News

"Some dentists recommend chlorella to patients who are having mercury amalgams removed." O'Brien 2001, Note cilantro has also found to be excellent at removing mercury. IMO use BOTH

"In the Kuopio Ischemic Heart Disease Risk Factor Study (KIHD), researchers from the University of Kuopio, Kuopio, Finland, noted that lipid peroxidation and excess risk of myocardial infarction (MI) could be best related to high mercury levels in the hair. High hair mercury levels were related to increased arterial wall thickness and growth in the carotid arteries. The team concluded: "Accumulation of mercury in the body is associated with accelerated atherosclerotic progression in men." Salonen et al. Circulation 1995;91:645-55

"Did this patient's health fail as a result of her new eating habits, or was it mere coincidence that her recent health decline followed the adoption of an extremely low-fat, low-cholesterol, low-animal-protein diet? I believe it was the latter and the scientific literature confirms my beliefs. The patient's reduction in cholesterol and total serum protein had made her vulnerable to bacterial and viral infection by promoting

T-cell suppression. This is especially so in the presence of mercury, which has been shown to reduce resistance to viruses, cancer and autoimmune disease. Low levels of cholesterol also make T-cell proliferation more difficult and the excretion of mercury nearly impossible." **Eric Davis, DDS**

"Most individuals can protect themselves against mercury by avoiding unnecessary exposure. That means using only composite dental fillings--never amalgam--and avoiding vaccines and pharmaceuticals that may contain thimerosal. Occasional fish consumption is fine in a healthy person who also consumes a diet rich in animal protein and fat, but tuna, swordfish and larger predatory species should be consumed only on rare occasions." **Eric Davis, DDS**

"Mercury is a fat soluble metal. That is, it is stored in the body's fat. The brain contains 60 percent fat and therefore is a common site for mercury storage." **Doctor Russell Blaylock MD**

Any science teacher encouraging students to put mercury in their mouths would be fired for gross negligence and likely prosecuted for endangering the health of a child. Yet dentists do it every day. And the US Food and Drug Administration lets them, all the while fully aware that there are serious safety concerns. At its website, FDA says, "Dental amalgams contain mercury, which may have neurotoxic effects on the nervous systems of developing children and fetuses. When amalgam fillings are placed in teeth or removed from teeth, they release mercury vapor. Mercury vapor is also released during chewing." (1) And a considerable amount is released, too. So-called "silver" fillings are 50% mercury."

Mercury Dental Amalgams Banned in 3 Countries

"What's more years of psychoanalysis will not reverse the depression that comes from profound omega 3 EFA deficiencies, a lack of vitamin B12, a low functioning thyroid or a chronic mercury toxicity." Doctor Mark Hyman MD
"Seven of the characteristic markers that we look for to distinguish Alzheimer's disease can be produced in normal brain tissues, or cultures of neurons, by the addition of extremely low levels of mercury." Doctor Boyd Haley, Professor, Chair Chem Dept, UK

"Selenium is an extremely low cost strategy that appropriately used may significantly help with many mercury issues." Doctor Gary Gordon MD

"I have worked with autistic children and have successfully co-developed a protocol getting mercury out of all children, without exception even when IV chelation previously had produced little or no effect. Sometimes we find mercury excretion levels off the chart. This heavy excretion sometimes continues for more than six months to two years." Doctor Gary Gordon MD

"Pure thimerosal is toxic at the low nanomolar level – an extremely low concentration, about 10,000 times less than the thimerosal concentration found in most vaccines." Doctor J. Curtis Pendergrass and Doctor Boyd Haley Kentucky University

"For fetuses, infants, and children, the primary health effect of methylmercury is impaired neurological development. Methylmercury exposure in the womb, which can result from a mother's consumption of fish and shellfish that contain methylmercury, can adversely affect a baby's growing brain and nervous system. Impacts on cognitive thinking, memory, attention, language, and fine motor and visual spatial skills have been seen in children exposed to methylmercury in the womb." EPA, Note they seem to be forgetting the thimerosal found in ALL vaccinations

"The Washington DOH routinely purchases mercury-free flu vaccines for infants under the age of three. However, pediatric patients aged 3-18 get flu vaccines that contain mercury. According to Secretary of Health, Mary Selecky, the rule will be suspended for six months, and applies only to vaccines against the swine flu. The Seattle Times quotes Selecky as saying that "the preservative, thimerosal, has never been linked to any health problems." The article goes on to state that a "vocal minority" believes the compound could be linked to autism. When a state Secretary of Health doesn't even acknowledge the truth (or is ignorant of the facts as they relate to health), you know something's seriously wrong. " Doctor Joseph Mercola MD
"To say that there is no evidence of link between thimerosal and biological damage is not a simple error or omission. It is an absolute lie, and you deserve better from your health officials." Doctor Joseph Mercola MD

"Researchers recently conducted a study on mercury intoxicated mice. The poison damaged several liver enzymes. The researchers gave the animals Tribulus terrestris extract at 6 mg/kg of body weight. The herb restored their liver enzymes to normal." Doctor Robert Rowen MD

"From a more mechanical perspective, mercury destroys the enzyme tubulin, which builds microtubules that play an important role in intracellular communication. According to Dr. Shade, you can clearly see how mercury stops the assembly of tubulin in neurological cells, causing them to fall apart instead." Mercury, Mercury Testing, and How to Detox Safely

"Mercury is in many of the foods we eat and it is also contained in a great many over-the-counter drugs and cosmetics; e.g. mascara, contact lens solution, hemorrhoid preparations, etc. The mercury ingredients used are thimerosal, phenylmercuric acetate,phenylmercuric nitrate, mercuric acetate, mercuric nitrate, MB for merbromin, and mercuric oxide yellow. Thus, sensitization to mercury can come from a number of sources." Mercury Filling Toxicity

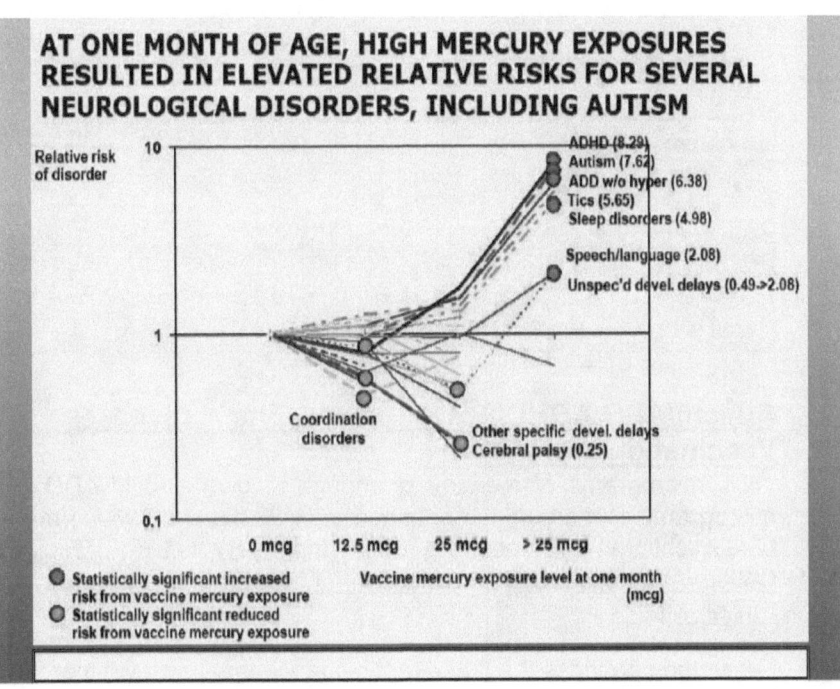

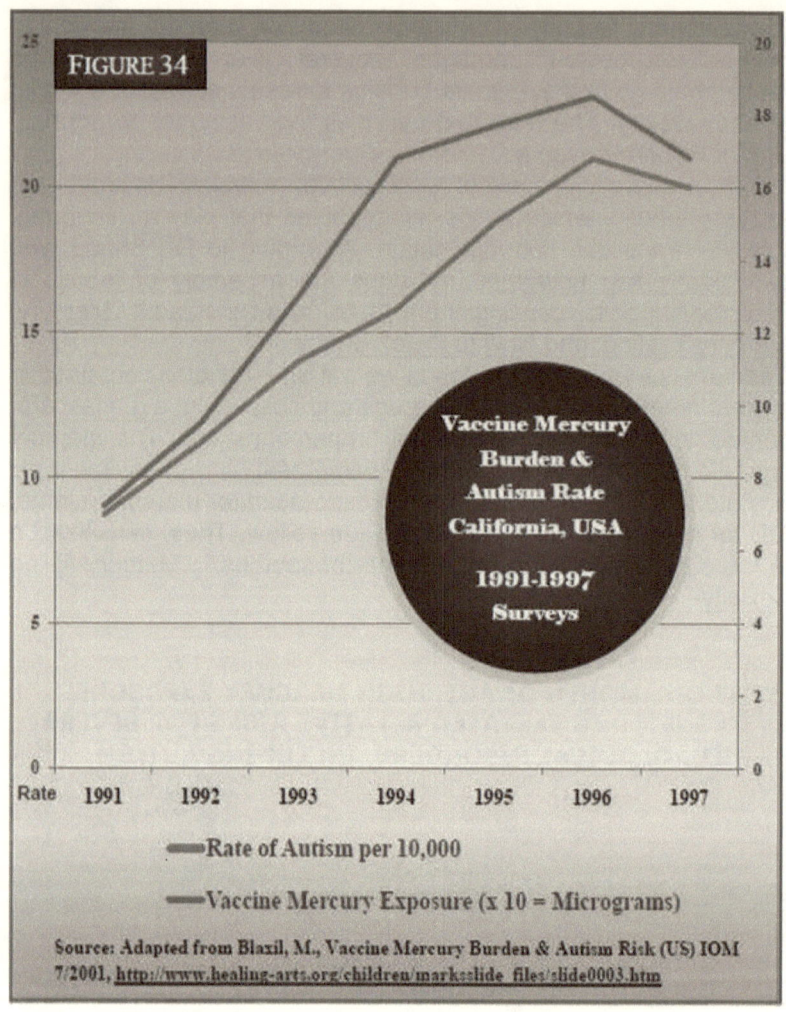

FIGURE 34

Rate of Autism per 10,000

Vaccine Mercury Exposure (x 10 = Micrograms)

Source: Adapted from Blaxil, M., Vaccine Mercury Burden & Autism Risk (US) IOM 7/2001, http://www.healing-arts.org/children/marksslide_files/slide0003.htm

Autism/ADD/ADHD/ASDs and The Diseases of Vaccination and Mercury

"12.5 micrograms of mercury is enough to give a child ADD. 25 micrograms of mercury is contained in ONE "flu shot" which many of our children are receiving each and every year." Vaccines Cause Autism, ADD/ADHD & Brain Damage - Dr. Kenneth P. Stoller, MD

"This study provides the first direct evidence that low level mercury exposure is indeed a participating factor that can initiate this

neurodegenerative process within the brain."   <u>Low Mercury</u>
<u>Causes Neurodegeneration (Brain Damage)</u>
<u>Mercury Pollution and Prenatal Exposure</u>

Doctor King Shows Decisively Thimerosal (Mercury) a
Strong Factor in Autism and the Diseases of Vaccination

"In New Delhi, India, prior to 2000, ASD/PDD (autism spectrum disorder/pervasive developmental disorder) symptoms were rare – typically only occurring in children who were vaccinated abroad. However, after the Indian pediatricians began recommending, in 2000, the addition of triple-dose Thimerosal-preserved Hib (Haemophilis influenza B) and Hep B (hepatitis B) vaccination programs to the existing Thimerosal-preserved triple dose DTP (diphtheria toxin, tetanus toxin and pertussis toxins) vaccination program recommended by the Government of India, the incidence of a childhood ASD/PDD diagnosis increased to 2 % to 4 % of vaccinated New Delhi children."   <u>Doctor Paul King PhD</u>
"To all but those who worship the god "vaccine" and knowingly sacrifice others children on this god's altar, these 9 Thimerosal-preserved vaccine doses are the cause of the chronic disease epidemic. After reading this declaration, this reporter challenges any one to say the cause is not the Thimerosal-preserved vaccines and/or that the epidemic of chronically ill children that has been engulfing us since the late 1980s and India since the early 2000s has no cause."   <u>Doctor Paul King PhD</u>

ALL "vaccines" contain mercury. **The average "mercury free vaccine" contains about 200 to 400 nanograms of mercury**. Mercury is used in the "vaccine" manufacturing process so even if they do not add mercury as an ingredient to the "vaccine" some mercury will end up in the "vaccine". If anyone gives you a hard time about this just ask them for a quantitative analysis to the nanogram level of the vaccine.
We thus see how the autism spectrum works. With a little mercury you may get ADD; add a little more mercury and you may get ADHD; add a little more mercury and you may get autism. IMO please get the mercury and "vaccinations" the hell out of your and your children's lives, because even if a "vaccine" contains zero nanograms of mercury it is greatly harming you and yours.

"The fetus doesn't have an immune system to protect it, and even small amounts of mercury will affect its development. ... There's no doubt that mercury-containing vaccinations have played a significant role here." Doctor Tom McGuire, DDS, book "Mercury Detoxification"

"Mercury can't be "broken" down, and the body must physically remove it. In the process, two molecules of glutathione are permanently lost for every atom of mercury that is removed and millions of atoms could be removed daily. This accounts for the depletion of glutathione" Doctor Tom McGuire, DDS, book "Mercury Detoxification"

"Mercury poisoning is a double-edged sword. The direct damage is caused when mercury attaches to proteins and enzymes (anything with a sulfhydryl group is fair game), altering their function and ultimately killing cells. Indirectly it can severely weaken the immune system and significantly deplete the body's most essential antioxidant, glutathione (GSH), weakening the body's ability to deal with other toxins and free radicals." Doctor Tom McGuire, DDS, book "Mercury Detoxification"

"Mercury is an extraordinarily poisonous substance in very small amounts. Even one atom will do some damage to the body." Doctor Tom McGuire, DDS, book "Mercury Detoxification"

The WHO has stated that there is no safe level of mercury and just one mcg of mercury contains approximately 4,300,000,000 atoms of it." Doctor Tom McGuire, DDS, book "Mercury Detoxification" Watch Mercury Ions Destroy Nerves

"Mercury has also the unfortunate ability to transfer from pregnant woman to their unborn babies. According to the Environmental Protection Agency, mercury passed on to the fetus during pregnancy may have lasting consequences, including memory impairment, diminished language skills and other cognitive complications." Dumbing Down Society Mercury in Foods and Vaccines

Book Review: Diagnosis Mercury

The author, Dr Jane Hightower is an internal medicine specialist working at a hospital in San Francisco, California. She receives a lot of referrals from primary care physicians and is known by her colleagues for her detective skills in identifying the cause of a variety of unusual or difficult disorders.

Mercury is a neurotoxic poison known far and wide for its health risks. IMO it is extremely important to have copious amounts of mercury chelators circulating and infused thought our bodies at all times. When a mercury atom enters our body it must be grabbed and ushered out immediately.

There is NO LEVEL of mercury in your body that is safe. One microgram is about 4.3 Billion atoms of mercury each of which can harm you.

Selenium (See Selenium Protocol) **also binds strongly with mercury protecting us from its damaging effects.** Mercury is excreted predominantly in the feces, but also in sweat and urine; thus your fetus Pregnancy Protection Protocol has NO way to eliminate mercury.

How Mercury Triggered The Age Of Autism
An Epidemic Induced By Added Doses of Thimerosal Preserved Vaccines
Blood mercury levels in autism spectrum disorder: Mercury as a Cause
Iatrogenic Death and Disease via Acute and Chronic Mercury Poisoning A Crisis in Medical and Dental Health

Mercury and Vaccination

© IStockphoto / Thinkstock

"Pure thimerosal is toxic at the low nanomolar level – an extremely low concentration, about 10,000 times less than the thimerosal concentration found in most vaccines."   Doctor J. Curtis Pendergrass and Doctor Boyd Haley Kentucky University

"A small dose of mercury that kills 1 in 100 rats and a dose of aluminum that will kill 1 in 100 rats, when combined have a striking effect: all the rats die. Doses of mercury that have a 1 percent mortality will have a 100 percent mortality rate if some aluminum is there. Vaccines contain aluminum."   Dr. Donald   W.   Miller,   Jr.,   Mercury   on   the   Mind, http://www.lewrockwell.com/miller/miller14.html, Note both mercury and aluminum are found in MANY "vaccines"

"We have found that clinically normal individuals aged 60-65 who receive influenza vaccine three or four times during a five-year period, will five years later have an incidence of Alzheimer's disease 10-fold greater than age-matched individuals who did not receive it."   **Doctor H. Hugh Fudenburg, MD**, Note and CDC cannot figure out where Alzheimer's is coming from. Right! 90% of our seniors are getting the "flu vaccine" toxin.

"If you get rid of the <u>*antiviral proteins in cells*</u>*, the [flu] virus can replicate 5 to 10 times faster. What that means is your cells have a mechanism that can block 80 to 90 percent of the [flu] virus that gets in. There is no better way to get rid of these protective types of proteins than injecting the body with more mercury."* Doctor Stephen Elledge Harvard Medical School

<u>**"Chronic mercury exposure is also a threat to our health and makes us especially vulnerable to flu infections**</u>. It has been shown that "prolonged exposure of mammals (white mice) to low mercury concentrations (0.008 – 0.02mg/m3) leads to a significant increase in the susceptibility of mice to pathological influenza virus strains. This is shown by more severe course of infection. In the experimental group more mice died (86 – 90.3 %) than in the unexposed animals (60.2 – 68 %), additionally the experimental group died more quickly. **The significant difference was in the appearance and degree of pneumonia in the effected animals**," <u>Doctor. I. M. Trakhtenbergin Chronic Effects of Mercury on Organisms</u>.[3], Note the H5N1 bio-weapon is a pathological influenza virus strain. Vitamin D in large amounts will help protect you from pneumonia.

"EVERYONE is ALLERGIC to thimerosal (mercury SALICYLATE). Salicylates provoke Serum Sickness. ... Allergy and Serum Sickness are 1 and 3 of the continuum of Hypersensitivity Reactions." Patrick Jordan
"This study provides the first direct evidence that low level mercury exposure is indeed a participating factor that can initiate this neurodegenerative process within the brain." Low Mercury Causes Neurodegeneration (Brain Damage)
"And how do we ever trust an organization like the FDA, which will not warn the public about the dangers of mercury containing dental amalgam no matter how much cancer, diabetes and neurological diseases the mercury fumes from the fillings leach directly into the lungs? This entire influenza, swine flu/respiratory pandemic is happening against a backdrop of an increasingly weakening population who become, in medical terms, accidents just waiting to happen. Respiratory distress is directly related to the entire spectrum of toxins whether heavy metal or complex chemicals fabricated for modern convenience. Pulmonary toxicity caused by acute exposure to mercury vapor has been demonstrated in animal studies.[4]" Mercury Contamination and the Flu
"The 1991 letter was from a Merck scientist, Dr. **Maurice Hilleman. In it, he informed the head of Merck's vaccine division that children were getting** a total dose of mercury 87 times higher than the **safe level set by the FDA at the time. (The FDA sets its safety limit higher than that of the EPA.)** After that memo, not only was no action taken to **lower the dose — but even more mercury containing vaccines were introduced, raising mercury levels to over 100 times FDA limits for toxicity and 150 times the EPA's safety level! But the FDA and the CDC, which were actively promoting the new vaccine schedule, did nothing."** Autism: The Silent Enemy Doctor Russell MD . Note this is an EXCELLENT treatise on many "vaccine" issues. You NEED to read it!
Doctor King Shows Decisively Thimerosal (Mercury) a Strong Factor in Autism and the Diseases of Vaccination
"In New Delhi, India, prior to 2000, ASD/PDD (autism spectrum disorder/pervasive developmental disorder) symptoms were rare – typically only occurring in children who were vaccinated abroad. However, after the Indian pediatricians began recommending, in 2000, the addition of triple-dose Thimerosal-preserved Hib (Haemophilis influenza B) and Hep B (hepatitis B) vaccination programs to the existing Thimerosal-preserved triple dose DTP (diphtheria toxin, tetanus toxin and pertussis toxins) vaccination

program recommended by the Government of India, the incidence of a childhood ASD/PDD diagnosis increased to 2 % to 4 % of vaccinated New Delhi children."   Doctor Paul King PhD
"To all but those who worship the god "vaccine" and nowingly sacrifice others children on this god's altar, these 9 Thimerosal-preserved vaccine doses are the cause of the chronic disease epidemic. After reading this declaration, this reporter challenges any one to say the cause is not the Thimerosal-preserved vaccines and/or that the epidemic of chronically ill children that has been engulfing us since the late 1980s and India since the early 2000s has no cause."   Doctor Paul King PhD

Mercury is found in all vaccines even the so called "Mercury Free Vaccines". If the vaccination dose contains less than one microgram of mercury the manufacturer can call his mercury an "impurity" and not explicitly list it. Please note that even one microgram of mercury is extremely toxic to the small undeveloped unprotected brain (See Brain Protection Protocol) of your fetus, infant or toddler. Many researchers believe without mercury many vaccines would not "work" even as poorly as they do. The mercury acts as an adjuvant IMO and is required as part of many vaccine's method of action.

It has been said that many sudden **deaths following vaccination (SIDs)** can be prevented by sufficient vitamin C and vitamin A reserves but we should probably also take a good look at magnesium (See Magnesium Protocol) as a crucial factor in sudden death after vaccination but if we keep injecting known toxins such as mercury and aluminum into our kids we should expect dead kids.

In adults it has been seen how high levels of mercury in the heart becomes dangerous. Since deficient magnesium and high levels of mercury are related; childhood vaccines would be especially dangerous to administer in children deficient in magnesium.

Mercury and MS and by extension IMO ALL Autoimmune Disease

"The oxygen binding sites in hemoglobin are a favorite of mercury. When enough mercury combines with the hemoglobin, the body experiences chronic fatigue due to lack of oxygen transport, and may create more red blood cells in compensation. This would show up as normal or high hemoglobin readings. Since the body cannot block the daily mercury doses released from amalgams, it will typically make more red blood cells to compensate for this daily contamination. Physicians can easily make the mistake of thinking that they couldn't possibly be hypoxic or anemic with normal hemoglobin. Once mercury is bound to hemoglobin, it will typically stay there for the lifetime of the red blood cell, which is approximately 120 days. Since one molecule of hemoglobin has four oxygen-binding sites, then one atom of mercury will drop the oxygen-carrying capacity of that hemoglobin molecule by 25% after binding. If two atoms of mercury attach, that hemoglobin molecule will have a 50% reduction of its oxygen carrying capacity, etc. After amalgam removal, the oxygen saturation in venous blood rises dramatically."   Mercury Filling Toxicity

"Mercury is implicated in metal-induced autoimmunity with the emphasis on multiple sclerosis (MS), rheumatoid arthritis (RA) and amyotrophic lateral sclerosis (ALS)..If everyone who had come down with MS, lupus, arthritis, epilepsy, leukemia, ALS, diabetes, etc., could relate their disease to dental procedures, the ensuing legal battle would be for more money than exists. A dentist can't legally throw amalgam material or extracted amalgam filled teeth in the trash, bury them in the ground, or put them in a landfill, but the ADA and the EPA say it's okay to put it in people's mouths." Mercury Filling Toxicity

"Evidence is mounting that low levels of magnesium contribute to the heavy metal deposition in the brain that precedes Parkinson´s, multiple sclerosis [Autism?] and Alzheimer´s. Research has shown that the symptoms of MS are very similar to Mercury poisoning.[2] Mercury is a primary cause of inflammation in our bodies."   MS: Magnesium, Selenium, Iodine and Mercury Connection

"Mercury is implicated in metal-induced autoimmunity with the emphasis on multiple sclerosis (MS), rheumatoid arthritis (RA) and amyotrophic lateral sclerosis (ALS)..If everyone who had come down with MS, lupus, arthritis, epilepsy, leukemia, ALS, diabetes, etc., could relate their disease to dental procedures, the ensuing legal battle would be for more money than exists. A dentist can't legally throw amalgam material or extracted amalgam filled teeth in the trash, bury them in the ground, or put them in a landfill, but the ADA and the EPA say it's okay to put it in people's mouths." Mercury Filling Toxicity

"This study provides the first direct evidence that low level mercury exposure is indeed a participating factor that can initiate this neurodegenerative process within the brain."   Low Mercury Causes Neurodegeneration (Brain Damage)

"Multiple Sclerosis patients have been found to have 8 times higher levels of mercury in the cerebrospinal fluid compared

to neurologically healthy controls. Inorganic mercury is capable of producing symptoms which are indistinguishable from those of multiple sclerosis." Doctor John Whitman Ray N.D., M.D. (M.A.)

There is absolutely no question that mercury toxicity can be an important factor in MS and by extension all auto-immune disease, IMO if you have a case of MS or any auto-immune disease that is not responding seriously consider removing every nanogram of mercury from your body, including any dental amalgams.

How Mercury toxicity is Expressed
"Nerve endings in the peripheral nervous system constantly scan their environment, engulfing foreign particles and bringing them across the cell membrane for inspection. These substances may then travel all the way up from the foot to the spinal cord to be presented to the nerve cells there. As it travels up the axon, mercury destroys a substance called tubulin, used as insulation for neurofibrils in the microtubules, effectively destroying the nerves. Within 24 hours of injecting a minute dose of mercury into a muscle anywhere in the body of test animals, it is detectable in the spinal cord and brain. The mercury is also found in the kidneys, lungs, bloodstream, connective tissue, adrenals and other endocrine glands. In the brain, it tends to congregate in the hypothalamus, which regulates the autonomic nervous system, and in the limbic system, believed to be the seat of emotions."
Mercury Filling Toxicity

- Altered cell membrane permeability
- Alteration of tertiary structure
- Alteration of enzyme function
- Interference in nerve impulses
- Alteration of the genetic code
- Inhibition of DNA repair
- Interference with endocrine function
- Contribution to autoimmune disease
- Digestion and absorption alteration
- Contribution to the development of antibiotic resistance

Autism and Mercury

"By the 1980s, my nieces and nephews received 8-9 vaccines when the U.S. resident population was 236 million people. By 1978, the rate of autism had increased four times, going from the previous rate of 1-in-10,000 to 1-in-2,500. Over the next ten years, the autism rate would climb again to 1-in-1,000 in 1991, when the DTP triple shot and hepatitis B were added to the vaccine chart, both of which contained thimerosal." **The Sting of Thimerosal in Autism**

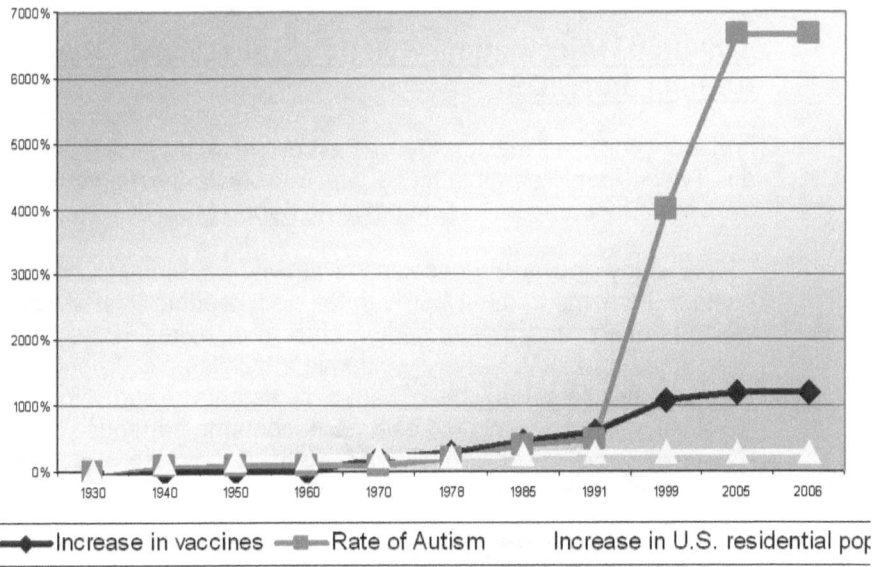

Aluminum Potentiates the Strongly Autism Associated Mercury

"A small dose of mercury that kills 1 in 100 rats and a dose of aluminum that will kill 1 in 100 rats, when combined have a striking effect: all the rats die. Doses of mercury that have a 1 percent mortality will have a 100 percent mortality rate if some aluminum is there. Vaccines contain aluminum." Doctor Donald W. Miller, Jr., Mercury on the Mind, http://www.lewrockwell.com/miller/miller14.html, Note both mercury and aluminum are found in MANY "vaccines"

"Two Neurotoxic Metals to Watch Out For: Mercury and Aluminum When it comes to maintaining brainpower, what you keep out of your body is as important as what you put into it. Minimizing your exposure to mercury is extremely important for your brain. <u>Dental amalgam</u> fillings are one of the worst sources of mercury. ... You may find my <u>mercury detox protocol</u> helpful. Also avoid <u>aluminum</u>, such as in antiperspirants, cookware, etc. Avoid flu vaccinations as they contain both mercury and aluminum. And stay away from all artificial sweeteners, such as <u>aspartame</u>, which are highly toxic to your brain."

Children, Vaccination, Tylenol, Glutathione, Mercury, Asthma and SIDS

"Tylenol promotes the toxicity of mercury. It is the worst thing to do. Tylenol depletes glutathione which protects the nervous cells from mercury damage."   <u>Doctor David Ayoub MD</u>

"Every study shows the association between acetaminophen and asthma, It seems to me it's time to tell people about this." Asthma, which is one of the most common chronic diseases in the United States, is a relatively new epidemic." <u>Children's Tylenol and related painkillers may be a cause of asthma surge</u>, Note this "relatively new epidemic" of asthma is contemporaneous with the "relatively new epidemic" of autism, which is itself contemporaneous with the "relatively new epidemic" of "vaccination"

<u>Empirical Data Confirm Autism Symptoms Related to Aluminum and Acetaminophen Exposure</u>

If you take the time to investigate these two sources and follow up links what you may find is this.

- Acetaminophen (Tylenol) depletes glutathione, this explains why acetaminophen can cause liver toxicity, the liver is the prime consumer of glutathione
- Glutathione is your bodies PRIME antioxidant
- Glutathione is found in HIGH concentration (normally) in your LUNGS
- Glutathione is REQUIRED to deal with mercury in your body
- Mercury RAPIDLY depletes glutathione

- When glutathione is depleted in the lungs it may bring on an asthma attack
- When glutathione is depleted in the brain breathing control centers (especially in a vaccinated child) it may bring on SIDS. You have to study Doctor Moulden a little to know this.
- When glutathione is depleted ANYWHERE in ANYONES body BAD things can happen.
- Please no acetaminophen (Tylenol) anytime for anything

Here is a dance of many partners, of disease and death, going on now in the USA. YOU can prohibit this disease and death in your family. All it takes is a little reading, viewing and action.

Doctor Boyd Haley discusses mercury toxicity and neurological diseases 2010
Doctor Boyd Haley mercury and oxidative stress IAOMT St. Louis 2011

"I think it's absolutely criminal to give mercury to an infant."
Doctor Boyd Haley PhD University of Kentucky

"The EPA and National Academy of Science reports that 8% to 10% of American women have such high body levels of mercury that would render increasingly susceptible to neurological damage any child they would give birth to."   Doctor Boyd Haley PhD

"Mercury can't be "broken" down, and the body must physically remove it. In the process, two molecules of glutathione are permanently lost for every atom of mercury that is removed and millions of atoms could be removed daily. This accounts for the depletion of glutathione" Doctor Tom McGuire, DDS, book "Mercury Detoxification"
"Mercury poisoning is a double-edged sword. The direct damage is caused when mercury attaches to proteins and enzymes (anything with a sulfhydryl group is fair game), altering their function and ultimately killing cells. Indirectly it can severely weaken the immune system and significantly deplete the body's most essential antioxidant, glutathione (GSH), weakening the body's ability to deal with other toxins and free radicals." Doctor Tom McGuire, DDS, book "Mercury Detoxification"
"The "dysfunctional mitochondria" can cause autism as they are producing more damaging free radicals than normal mitochondria. Mitochondria made toxic by ethylmercury, released by thimerosal,

would do the same thing as the tightest binding site for mercury in the body is in the "iron sulfur centers" of the electron transport chain that carries the 'electron' used normally to produce the mitochondrial pH gradient that makes energy in the form of ATP. Ethylmercury blocking of this transport chain leads to the production of hydroxyl free radicals by release of the electron to react with water. This causes oxidative stress and low reduced glutathione levels as observed in autism and many neurological illnesses."   Doctor Boyd Haley PhD

Doctor Boyd Haley has done extensive and excellent work in the field of mercury toxicity. If you but listen to the videos above you will learn a lot about mercury toxicity and how a single atom of mercury can kill energy transport within the cell.

Application of Doctor Haley's Mercury Findings to Remove Mercury
"One sulfur-based food is garlic, which has gained a reputation for being good for mercury detoxification. As explained by Dr. Shade, garlic contains sulfur molecules that trigger the translocation of the Nrf2 proteins into your cells. Thus they trigger an upregulation of your glutathione system. So it's really your glutathione system as a whole that is responsible for the chelating effect, but the sulfur compounds in the garlic upregulate that system, allowing it to do its job."   *Mercury, Mercury Testing, and How to Detox Safely*

- Glutathione (GSH) **is the ONLY way to remove mercury from your body**
- Test for GSH levels
- Your goal is to replete low GSH to normal levels and above
- Both for prevention and treatment of autism we must MAXIMIZE GSH
- Whey helps increase GSH
- High ORAC antioxidants, spices, food and supplements helps increase available GSH
- Sulfur helps increase GSH
- Anything you can do to take other responsibilities from GSH will free GSH to transport mercury out of body
- Alpha Lipoic Acid helps increase available GSH
- To battle mercury in the mitochondria SOD helps free up GSH

Doctor Haley tells us the way that mercury is usually removed from our bodies is that two molecules of glutathione bind an atom of mercury and flush it into the body fluid where it is removed into the urine by the kidneys. Therefore we must maximize the glutathione available to escort the mercury out of our bodies. We can do this by taking the maximum of antioxidants to take some of the load off glutathione' antioxidant responsibilities and increasing our production of glutathione.

Doctor Boyd Haley's OSR Mercury Chelator Note Doctor Haley has removed this product from the market. IMO it may have worked too well. This powerful detoxifier probably overloaded the first phase of the detoxification process. If you have some or can get some. Start LOW go SLOW. Start with far less than a gram IMO. I have no doubt it is a superior antioxidant.

Eating Fish can be Better than a Visit to Your Dentist or Pediatrician

"When selenium, which is present in most fish, and mercury are found together, they connect forming a new compound. This makes it difficult for the human body to absorb the mercury separately. As we shall also see scientists have also tagged cysteine in fish binding with mercury also making it safer to eat. When mercury "binds" to selenium or cysteine it is no longer free to "bind" to anything else — like brain or kidney tissue. An accurate picture of the health consequences of eating fish must include other substances that affect the way mercury interacts with the human body." Mercury Causes Chronic Disease
Selenium protects you from the mercury in fish somewhat and may also provide some protection from your ped and dentist.

Avoid Mercury

"High-fructose corn syrup (HFCS) is known to be a contributor to obesity and a number of serious health ailments, but very few health experts and medical publications ever discuss the highly toxic substance hiding within a large percentage of HFCS — mercury. Found in a large number of processed foods and sodas, mercury-containing HFCS is toxic in all forms and may be contributing to the rampant disease rates inside the United States and elsewhere." Mercury in HFCS

"A second mechanism of producing neurodegenerative diseases is even more impressive, called excitotoxicity. Excitotoxicity, a mechanism by which excess glutamate accumulates outside the neuron, thereby leading to death of the cell by an excitation process, has been linked to mercury neurotoxicity as early as 1993. More recent studies have confirmed this mechanism and clearly demonstrate, even in concentrations below that known to cause cell injury; mercury can paralyze the glutamate removal mechanism, leading to significant damage to synapses, dendrites and neurons themselves." Mercury Filling Toxicity, Note excitotoxicity is also caused by aspartame, **NutraSweet®**

"The oxygen binding sites in hemoglobin are a favorite of mercury. When enough mercury combines with the hemoglobin, the body experiences chronic fatigue due to lack of oxygen transport, and may create more red blood cells in compensation. This would show up as normal or high hemoglobin readings. Since the body cannot block the daily mercury doses released from amalgams, it will typically make more red blood cells to compensate for this daily contamination. Physicians can easily make the mistake of thinking that they couldn't possibly be hypoxic or anemic with normal hemoglobin. Once mercury is bound to hemoglobin, it will typically stay there for the lifetime of the red blood cell, which is approximately 120 days. Since one molecule of hemoglobin has four oxygen-binding sites, then one atom of mercury will drop the oxygen-carrying capacity of that hemoglobin molecule by 25% after binding. If two atoms of mercury attach, that hemoglobin molecule will have a 50% reduction of its oxygen carrying capacity, etc. After amalgam removal, the oxygen saturation in venous blood rises dramatically."

Kids Exposed to Mercury, Lead at Risk for ADHD

Most individuals can protect themselves against mercury by avoiding unnecessary exposure. That means using only composite dental fillings--never amalgam--and ALWAYS avoiding ALL vaccines, flu shots and pharmaceuticals. Wild deep ocean fish consumption is fine in a healthy person who also consumes a diet rich in cholesterol, animal protein and fat, but tuna, swordfish and larger predatory species should be consumed only on rare occasions or not at all.

Mercury and Dental fillings

A "Silver" tooth filling amalgam is 50% mercury and 25% silver. What does this tell you?

One group that is particularly badly treated are children with disabilities... [T]here was just an all-out battle in Philadelphia, because we succeeded in getting a fact sheet law, so the parents were reading the fact sheets, saying, "I don't want amalgam." The dentists serving children with disabilities were telling the parents, "You will get the filling I decide on"... Parents were forced to leave the office or accept a mercury filling! These dentists were backed up by the Pennsylvania Dental Association. That was condemnable. The ADA in fact issued an apology recently for its history of racism. That appears to continue with their attitude that those who are disabled have no rights to mercury-free dentistry. That's one of the battles that we're [facing]... to protect those who are less able to fend for themselves in this economic society."
Charlie Brown LdL

"Mercury is a multipotent cytotoxin that intervenes in the primary processes of the cell by bonding strongly with sulfhydryl and selenohydryl groups on albumen molecules in cell membranes, receptors and intracellular signal links, and by modifying the tertiary structure. The structure of albumen molecules is genetically determined, and this leaves ample scope for genetic polymorphism to manifest itself in varying sensitivity and types of reaction to mercury exposure. Mercury is toxic because it induces production of free oxygen radicals and modifies the redox potential of the cell."

"Research from Sweden has demonstrated that removal of dental amalgams from about 700 subjects with neurological problems led to clinical improvements in about 70% of the subjects, along with a significant drop in the blood mercury levels of the subjects. There are also reports that individuals with multiple sclerosis had less deleterious events when their amalgams were removed. Research has shown that individuals who died of idiopathic dilated cardiomyopathy have 20,000 times more mercury in their heart tissue that found in other forms of heart disease. This was published in a 1999 issue of the Journal of American Cardiology. Yet no NIH grants or programs have been developed to pursue this lead"

"Further, electron microscopy of dental amalgams clearly shows droplets of mercury liquid in dental amalgam pores. Heating the amalgam releases this mercury quickly and causes the droplets to disappear. A massive German university study found toxic levels of mercury in the saliva of several thousands of subjects and the amount was correlated to dental amalgams. There is no scientific controversy about the nature and amount of mercury being emitted from a dental amalgam. The only controversy is maintained by the inaccurate and manipulated data (as well as Congressional lobbying efforts) put forth by the pro-amalgam elements in organized dentistry, including the dental branch of the FDA and the NIDCR."

"The half life of mercury vapor in the urine and blood is very short and such levels are not a good measure of exposure. Many acutely exposed individuals will have urine levels considered non-toxic, yet have high mercury levels in their organs years later when they die. Most studies on children indicate that the ones with the highest urine, blood or hair levels of mercury were the healthiest. That is because of those exposed to mercury; the ones with the highest urine, blood and hair levels are the ones effectively excreting the mercury. Three different research groups have shown that autistic children have much lower mercury in their hair, yet have higher body burdens of mercury. This implies that an inability to excrete mercury by a subset of the population represents those that will respond badly to a low chronic exposure to mercury."

"Amalgam fillings contain approximately 50% mercury, 30% copper, 14% each of tin and silver, and 1% zinc. All five metals in amalgam fillings are toxic. These metals react with each other and form sixteen more corrosion products, all of which are toxic. The continued use of mercury amalgam restorations has spawned whole industries whose livelihoods are dependent upon removal of this toxic metal from dental office waste streams where amalgam restorations are both placed and removed."

"Mercury is implicated in metal-induced autoimmunity with the emphasis on multiple sclerosis (MS), rheumatoid arthritis (RA) and amyotrophic lateral sclerosis (ALS)..If everyone who had come down with MS, lupus, arthritis, epilepsy, leukemia, ALS, diabetes, etc., could relate their disease to dental procedures, the ensuing legal battle would be for more money than exists. A dentist can't legally throw amalgam material or extracted amalgam filled teeth in the trash, bury them in the ground, or put them in a landfill, but the ADA and the EPA say it's okay to put it in people's mouths."

"Consumers aren't being told the truth, that amalgam fillings contain 50% mercury, a known neuro toxin. Worse, they are deceived: the ADA still uses the deceptive word "silver" to describe a product that is mainly mercury, thus hiding the product's main ingredient. The ADA has a "gag rule" and enforces it through state dental boards, which prohibits dentists from initiating discussion critical of amalgam's health effects. Substitutes exist for amalgam, including composite (or resin), ceramic, porcelain and gold. Because of the slightly higher cost of placing composites, the most commonly used alternative dental filling, Medicaid and barebones insurance plans force children to use amalgam, even though it is well known that some will have adverse reactions."

"Based on the known toxic potential of mercury and its documented release from dental amalgams, usage of mercury containing amalgam increases the health risk of the patients, the dentists and the dental personnel." International Conference on Biocompatibility of Materials 1988

I had a mouth full of mercury, mostly because I did not start taking care of my teeth until I was about 15. For the last fifteen or so years I have gotten my fillings made with composite. Mercury is still leaking into my body every time I chew or bite. I would be worried and have my fillings removed, ouch, if it were not for Mercury Chelators. I take so many Mercury Chelators on a regular basis I doubt there is a mercury molecule in my body other than in my teeth. A <u>chelator grabs on to its object</u> so your liver can eliminate the duo.
<u>Smoking Teeth = Mercury Vapors</u>

An Environmental Disaster in the Making

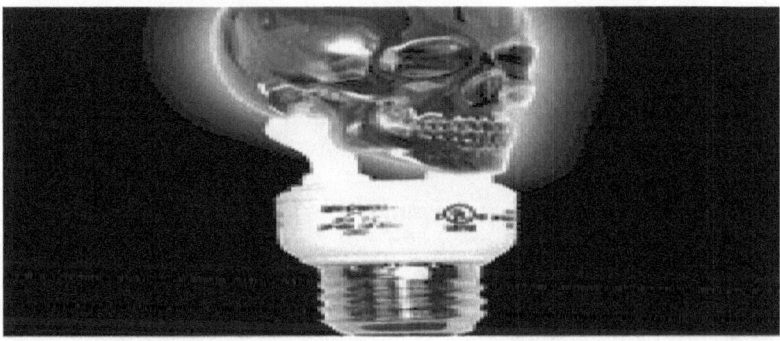

"CFL bulbs are filled with mercury. They'll tell you it's a "tiny" amount, but a single bulb contains enough mercury to contaminate 6,000 gallons of water. There's emerging evidence that these bulbs can cause seizures and dizziness, worsen migraine headaches and send lupus patients doubling over in agony. They've been linked to skin disorders and even cancer. I can only wonder what else they'll discover when hundreds of millions of people are forced to sit under these bulbs all day, every day... at home, at work, and everywhere in between. Light bulbs break -- and it's just a matter of time before you and your family are exposed to mercury from a shattered CFL bulb. What's going to happen when all the mercury from discarded and "recycled" bulbs seeps into the ground and ends up in the already-filthy water supply."   Doctor William Douglass MD

Toxic Light - The Dark Side of Energy Saving Bulbs

 Folks this is guaranteed to be a disaster IMO. The nerve, arrogance and pigheaded gall of our "government" is beyond limit. I have a reserve of 100 incandescent light bulbs.

Compact fluorescent bulbs release cancer-causing chemicals when turned on, says new research

The practical questions each parent and family needs to face in regards to mercury are:
1) Should I let dentists put mercury in my children's mouths?
2) Should I let the pediatrician inject mercury into my baby's body?   MANY "vaccines" contain mercury TODAY*
3) Should I eat fish that contain high levels of mercury?
4) Should I live anywhere near a coal fired power station or...
5) Should I throw out my old mercury thermometer?
6) Is mercury involved in the etiology of my or my children's diseases and learning disorders?
* Even many so called "mercury free vaccines" are manufactured by a mercury process; thus most childhood disease "vaccines" contain about 300 nanograms of mercury

How Much Mercury do you want
"I'll tell you that I like people [to be] under 0.05 nanograms per milligram for the dental form of mercury. That's when you know that you're really cleaned up. For methylmercury, that number can be a little higher. Depending upon the state of their health, I like them down below 2 nanograms per liter, or if they have shown some sensitivity to mercury, then I like them under 1 nanogram per liter." Mercury, Mercury Testing, and How to Detox Safely

Your body is composed of about 80 liters of matter. So you should want about 80 nanograms or less of mercury in your body. There are about 25,000 nanograms of mercury in a "flu vaccination"; all of this goes directly into your body. Most childhood "vaccines" contain about 300 nanograms or much more of mercury; all of this goes directly into your body. The amount of mercury vapor a "silver filling" releases per day is about 10,000 nanograms; some of this gets into your body. You decide.
http://www.biodentistrydrvizcarra.com/media/MERCURY_FILLING S_TOXICITY.pdf

Mercury Detoxification
"According to Dr. Shade, effective detoxification is highly dependent on your glutathione and sulfur metabolism. For example, when your glutathione and sulfur levels increase, you'll typically see higher levels of mercury coming out of your hair. It's important to realize, however, that detoxing cannot be achieved overnight. Or even in a few weeks. According to Dr. Shade, most people will typically need an entire year to detox, depending on how sick you are. The sicker you are, the longer it will take, simply because you have to go slower when you're ill." Mercury, Mercury Testing, and How to Detox Safely

"But then you have to keep in mind one of the aspects of detoxification. You got to be mindful of that, and incorporate a lot of this into your life. I mean a lot of it is lifestyle changes that are going to keep you effectively detoxifying." Mercury, Mercury Testing, and How to Detox Safely

"The core of the detoxification system is the glutathione system. Notice that I don't say, "it's glutathione." No, it's the glutathione system. Glutathione binds to metals and can move them out of your body, but it doesn't do that alone." Mercury, Mercury Testing, and How to Detox Safely

"If we look at what the requirements for resistance to metals and effective detoxification (meaning not only the resistance but shuttling them out), you need healthy levels of glutathione in your cells. You [also] need activity of... a phase II detoxification enzyme called glutathione S-transferase." Mercury, Mercury Testing, and How to Detox Safely

"We use a lot of this liposomal glutathione. We have recently taken on the manufacturing of that that's making a very small, high integrity liposome. But aside from liposomes, whey protein is my favorite way to get those precursors into the body." Mercury, Mercury Testing, and How to Detox Safely

"According to Dr. Shade, if your body is making glutathione well, then just taking whey protein should be sufficient. But if your detoxification system is severely compromised, the liposomal glutathione can serve an important function as it can offer immediate relief against the oxidative stress caused by the mercury. He typically recommends using both for people who are ill." *Mercury, Mercury Testing, and How to Detox Safely*

"It's an interesting fact that some people with high mercury exposure don't become toxic, yet others with relatively low exposure do. Why is this? Why does one person get really sick from her amalgams while another is perfectly fine? The difference lies in your ability to detoxify naturally. You already have a system in place for removing mercury and other heavy metals from your body. Mercury's half-life can range from 40 to 120 days, and the faster you can clear it out, the less you'll be affected." Revised Protocol for Detoxifying Your Body from Mercury **Exposure**

"It's important to realize that people are either slow detoxifiers or fast detoxifiers, and a small genetic subset are *super slow* detoxifiers. If you are in the super slow group, your detox system is significantly impaired and the result can be mercury overload. How quickly you detoxify on your own depends on a several factors, such as your exposure level, genetic makeup, genetic expression, and overall health. For example, if your progesterone levels are low, you can't detoxify as well, and unfortunately, decreased progesterone levels are common today." Revised Protocol for Detoxifying Your Body from Mercury **Exposure**

"It is important to realize that any mercury detoxification is a marathon and NOT a sprint. You do NOT want to do this quickly. Even if you believe you are healthy you want to start this process SLOWLY as you could easily cause severe flare ups. I am one of the healthiest people I know and when I did my program I did it over six months. Some people may need to do it far more slowly and may need a few years to effectively eliminate the mercury safely." What is Chlorella Good For?

The above papers and videos will go a long way to helping you understand the big mercury detoxification picture.

Required for Mercury Detoxification

- Glutathione
- Sulfur
- Mercury Chelators
- Glutathione System

- Glutathione Precursors
- Whey
- Cysteine
- NAC
- Garlic
- Glutathione glutathione S-transferase
- Vitamin C to recharge the antioxidant system
- One year or more   for serious cases
- Liposomal Glutathione   for serious cases

Mercury Chelators

"Activated charcoal is the single best supplement for enhancing detoxification. *You can take charcoal to wipe out decades of toxic heavy metals that may have been building up in your body. Harmful metals like arsenic, copper, mercury, and lead. And it could one day save your life. An activated charcoal detox leaves you feeling like a new person – pumped up, recharged, and bursting with energy. As if you were suddenly 20 years younger. It's best to use a powder form, mixed into a liquid. Tablets or capsules take too long to absorb and release the activated charcoal. And the dose is usually too small to do the job. Take 20-30 grams a day of powdered activated charcoal (in divided doses) mixed with water over a period of 1-2 weeks."*   Doctor Al Sears MD

"Selenium is an extremely low cost strategy that appropriately used may significantly help with many mercury issues." Doctor Gary Gordon MD

"Mercury occurs naturally in almost all foods but within very small amounts." Doctor Anthony Cichoke, Note this demonstrates your body can normally deal with nanograms of mercury in your digestive system by for instance chelating it

"The good news is that mercury toxicity is reversible. They can pull mercury out to the body with Chelation agents; they can do this nutritionally; they can turn on the bio-chemical pathways that are plugged up. So it is reversible and there are a lot of ways to do it." Doctor David Ayoub MD

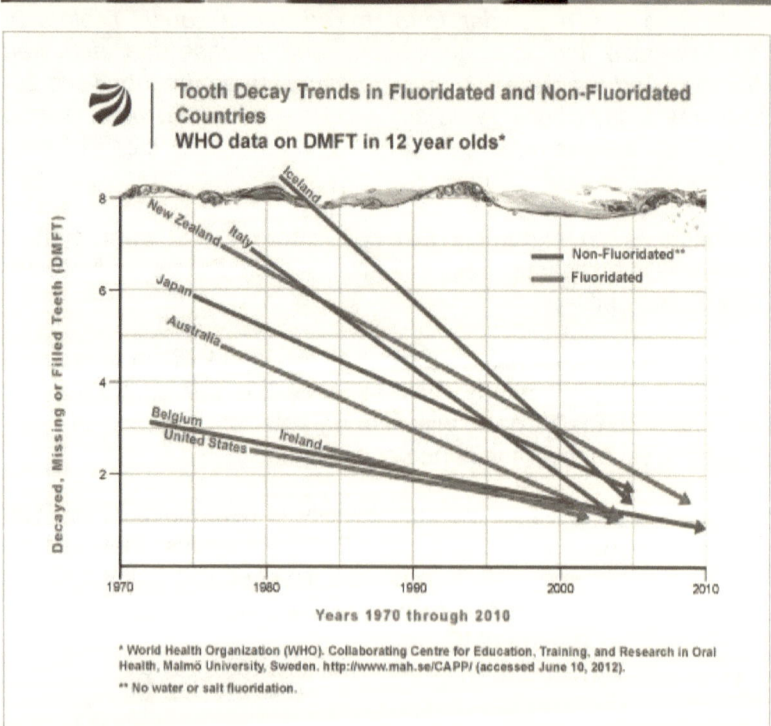

Tooth Decay Trends in Fluoridated and Non-Fluoridated Countries
WHO data on DMFT in 12 year olds*

* World Health Organization (WHO). Collaborating Centre for Education, Training, and Research in Oral Health, Malmö University, Sweden. http://www.mah.se/CAPP/ (accessed June 10, 2012).

** No water or salt fluoridation.

U.S. water fluoridation began in 1945, never FDA approved, yet continues today

History of Medicine Fact #8: U.S. water fluoridation began in 1945 and continues today, despite the fact that the FDA has never approved it

S. D. Wells
Natural News
June 24, 2012

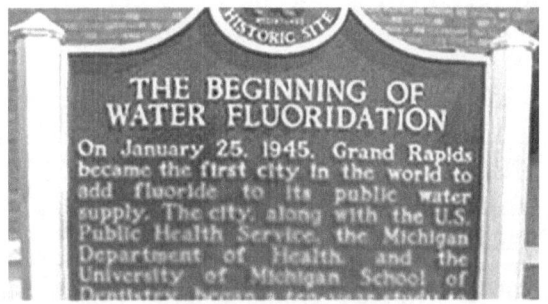

(NaturalNews) **The year before water fluoridation began in the United States, the entire dental profession recognized that fluoride was detrimental to dental health. In fact, in 1944 the*Journal of the American Dental Association* reported that using between 1.6 and 4 ppm (parts per million) fluoride in water would cause 50% of adults to need false teeth. On top of that, the world's largest study looked at 400,000 students, revealing that tooth decay increased in over 25% with just 1ppm fluoride in drinking water. (http://www.healthy-communications.com)**

Yet still, in 1945, fluoride was put into municipal water systems in Newburgh, New York, and Grand Rapids, Michigan. Over the next 50 years, more than 60 percent of the U.S. population was "fluoridated" at a minimum of 1 ppm. Currently, over 75% of the United States **water supply contains this deadly toxin.**

One part of the hoax, "fluoride helps with tooth formation," was removed from the "American Fluoride Campaign" early on. Realizing this might expose the entire campaign as fraudulent, the FDA and CDC simply

removed that language, but kept the masses believing that fluoride keeps dental cavities at bay.
Over 70% of America still clings to the multi-faceted myth

Research proves that fluoride is an extremely neurotoxic chemical which interrupts basic functions of nerve cells in the brain and **can lead to Alzheimer's, atherosclerosis (hardening of the arteries), infertility, birth defects, diabetes, cancer and lowered IQ.** The aluminum "tricks" the blood-brain barrier and allows chemical access to brain tissue.

<u>**Think fluoride is used by the rest of the world? France, Germany, Japan, Sweden, Denmark, Holland, Finland, India and Great Britain have all rejected its use after special commissions and health secretaries reviewed the negative evidence.**</u>

Think fluoride cleanses the water? Fluoride is one of the basic ingredients in military nerve gas. Sodium fluoride is a hazardous waste by-product from the manufacture of aluminum and fertilizer, and it is a common ingredient in roach and rat poisons.

Think fluoride fights cavities and strengthens bones? Dental fluorosis is often caused by over-exposure to fluoride when the dental enamel is mineralizing during childhood. Fluoride is unique in its ability among acids to penetrate tissue, *causing soft tissue damage and bone erosion* as it leaches calcium and magnesium from the body. (http://tuberose.com/Fluoride.html)

Think fluoride evaporates from water? Fluoride does not evaporate from water left sitting out. Also, boiling or freezing won't help at all, and basic filters like Brita do not remove it. Reverse osmosis does remove it, and **natural spring water does not contain it.**

<u>**Because the ADA maintains a stranglehold on the dental profession, no dentists are ever openly critical of**</u>

<u>fluoride. The ADA can influence State Dental Boards which can take away a dentist's license, so you won't hear anything negative about it from your dentist.</u> **Most brands of toothpaste contain at least 1,000 ppm fluoride, so if a child were to eat an entire tube, he/she would die.**

Fluoride has never received FDA approval and does not meet "requirements of safety and effectiveness." **The FDA states that fluoride is a prescription drug. Because this "drug" is put in municipal water,** *there is absolutely no control over individual dosage.*

So, why on earth would the USA's regulatory agencies allow such a nightmare to perpetuate? In the early 1900's, when important vitamins (like B12) were discovered and natural remedies became popular, medicine was basically unprofitable. Fluoridation was a **planned experiment of mass medication** to induce diseases that would later be "treated" with expensive healthcare, and that is why government paid healthcare in America is nothing but a pipe dream.

Learn more about the true history of modern medicine in this revealing report. Free downloadable PDF. Covers Nazi connections with Big Pharma, war crimes of Bayer, the weapons origins of prescription medications, the shocking history of psychiatric medicine and much more.

Sources for this article include:
http://www.fluoridedebate.com/
http://www.healthy-communications.com
http://www.cancer.gov/cancertopics/factsheet/Risk/fluoridated-water
http://www.nap.edu/openbook.php?record_id=2204&page=R1
http://www.consumerhealth.org/articles/display.cfm?ID=199903
03222823
http://tuberose.com/Fluoride.html
http://www.scienceclarified.com/Ga-He/Halogens.html
http://www.chemistryexplained.com/elements/A-C/Chlorine.html

This article was posted: Sunday, June 24, 2012 at 2:40 pm

The Fluoride Conspiracy

"Tell a lie loud enough and long enough and people will believe it."
- Adolf Hitler

"Fluoridation is the greatest case of scientific fraud of this century."
- Robert Carlton, Ph.D, former EPA scientist, 1992

The history of forcing fluoride on humans through the fluoridation of drinking water is wrought with lies, greed and deception. Governments that add fluoride to drinking water supplies insist that it is safe, beneficial and necessary, however, scientific evidence shows that fluoride is not safe to ingest and areas that fluoridate their drinking water supplies have higher rates of cavities, cancer, dental fluorosis, osteoporosis and other health problems. Because of the push from the aluminum industry, pharmaceutical companies and weapons manufacturers, fluoride continues to be added to water supplies all over North America and due to recent legal actions against water companies that fluoridate drinking water supplies, precedent has been set that will make it impossible for suits to be filed against water suppliers that fluoridate. There is a growing resistance against adding toxic fluoride to our water supplies, but unfortunately, because fluoride has become "the lifeblood of the modern industrial economy"(Bryson 2004), there is too much money at stake for those who endorse water fluoridation . The lies of the benefits of water fluoridation will continue to be fed to the public, not to encourage health benefits to a large number of people, but to profit the military-industrial complex.

The story begins in 1924, when Interessen Gemeinschaft Farben (I.G. Farben), a German chemical manufacturing company, began receiving loans from American bankers, gradually leading to the creation of the huge I.G. Farben cartel. In 1928 Henry Ford and American Standard Oil Company (The Rockefellers) merged their assets with I.G. Farben, and by the early thirties, there were more than a hundred American corporations which had subsidiaries and co-operative understandings in Germany. The I.G. Farben assets in America were controlled by a holding Company, American I.G. Farben, which listed on it's board of directors: Edsel Ford, President of the Ford Motor Company, Chas. E. Mitchell, President of Rockerfeller's National City Bank of New York, Walter Teagle,

President of Standard Oil New York, Paul Warburg, Chairman of the federal reserve and brother of Max Warburg, financier of Germany's War effort, Herman Metz, a director of the Bank of Manhattan, controlled by the Warburgs, and a number of other members, three of which were tried and convicted as German war criminals for their crimes against humanity. In 1939 under the Alted agreement, the American Aluminum Company (ALCOA), then the worlds largest producer of sodium fluoride, and the Dow Chemical Company transferred its technology to Germany. Colgate, Kellogg, Dupont and many other companies eventually signed cartel agreements with I.G. Farben, creating a powerful lobby group accurately dubbed "the fluoride mafia"(Stephen 1995).

At the end of World War II, the US government sent Charles Eliot Perkins, a research worker in chemistry, biochemistry, physiology and pathology, to take charge of the vast Farben chemical plants in Germany. The German chemists told Perkins of a scheme which they had devised during the war and had been adapted by the German General Staff. The German chemists explained of their attempt to control the population in any given area through the mass medication of drinking water with sodium fluoride, a tactic used in German and Russian prisoner of war camps to make the prisoners "stupid and docile"(Stephen 1995). Farben had developed plans during the war to fluoridate the occupied countries because it was found that fluoridation caused slight damage to a specific part of the brain, making it more difficult for the person affected to defend his freedom and causing the individual to become more docile towards authority. Fluoride remains one of the strongest anti-psychotic substances known, and is contained in twenty-five percent of the major tranquilizers. It may not seem surprising that Hitler's regime practiced the concept of mind control through chemical means, but the American military continued Nazi research, exploring techniques to incapacitate an enemy or medicate an entire nation. As stated in the Rockerfeller Report, a Presidential briefing on CIA activities, "the drug program was part of a much larger CIA program to study possible means of controlling human behavior"(Stephen 1995).

The 'dental caries prevention myth' associated with fluoride, originated in the United States in 1939, when a scientist named Gerald J. Cox, employed by ALCOA, the largest producer of toxic fluoride waste and at the time being threatened by fluoride damage claims, fluoridated some lab rats, concluded that fluoride reduced cavities and claimed that it should be added to the nation's water supplies. In 1947, Oscar R. Ewing, a long time ALCOA lawyer, was appointed head of the Federal Security Agency , a position that placed him in charge of the Public Health Service(PHS). Over the next three years, eighty-seven new American cities began fluoridating their water, including the control city in a water fluoridation study in Michigan, thus eliminating the most scientifically objective test of safety and benefit before it was ever completed.

American 'education and research' was funded by the Aluminum Manufacturing, Fertilizer and Weapons Industry looking for an outlet for the increasingly mounting fluoride industrial waste while attaining positive profit increase. The 'discovery' that fluoride benefited teeth, was paid for by industry that needed to be able to defend "lawsuits from workers and communities poisoned by industrial fluoride emissions" (Bryson 1995) and turn a liability into an asset. Fluoride, a waste constituent in the manufacturing processes of explosives, fertilizers and other 'necessities', was expensive to dispose of properly and until a 'use' was found for it in America's water supplies, the substance was only considered a toxic, hazardous waste. Through sly public re-education, fluoride, once a waste product, became the active ingredient in fluorinated pesticides, fungicides, rodenticides, anesthetics, tranquilizers, fluorinated pharmaceuticals, and a number of industrial and domestic products, fluorinated dental gels, rinses and toothpastes. Fluoride is so much a part of a multibillion-dollar industrial and pharmaceutical income, that any withdrawal of support from pro-fluoridationists is financially impossible, legally unthinkable and potentially devastating for their career and reputation.

Funded by US industrialists, in an attempt to encourage public acceptance of fluoride, Edward Bernays, known also as the father of PR, or the original spin doctor, began a campaign of deception to persuade public opinion. Barnays explained "you can get practically any idea accepted if doctors are in favour. The public is willing to accept it because a doctor is an authority to most people, regardless of how much he knows or doesn't know"(Bryson 2004). Doctors who endorsed fluoridation didn't know that research discrediting fluoride's safety was either suppressed or not conducted in the first place. Fluoride became equated with scientific progress and since it was introduced to the public as a health-enhancing substance, added to the environment for the children's sake, those opposing fluoride were dismissed as cranks, quacks and lunatics. Fluoride became impervious to criticism because of a relentless PR offensive, but also because of it's overall toxicity. Unlike chemicals that have a signature effect, fluoride, a systemic poison, produces a range of health problems, so it's effects are more difficult to diagnose.

Recently declassified US Military documents such as Manhattan Project, shows how Fluoride is the key chemical in atomic bomb production and millions of tonnes of it were needed for the manufacture of bomb-grade uranium and plutonium. Fluoride poisoning, not radiation poisoning, emerged as the leading chemical health hazard for both workers and nearby communities. A-bomb scientists were ordered to provide evidence useful for defense in litigation, so they began secretly testing fluoride on unsuspecting hospital patients and indignant, mentally retarded children.. "The August 1948 Journal of the American Dental Association shows that evidence of adverse effects from fluoride was censored by the US Atomic Energy Commission for reasons of "national security" (Griffiths 1998). The only report released stated that fluoride was safe for humans in small doses.

During the Cold War, Dr. Harold C. Hodge, who had been the toxicologist for the US Army Manhattan Project, was the leading scientific promoter of water fluoridation. While Dr. Hodge was reassuring congress of the safety of water fluoridation, he was covertly conducting one of the nation's first public water fluoridation experiments in Newburgh, New York, secretly studying biological samples from Newburgh citizens at his US laboratory at the University of Rochester. Since there are no legal constraints against the suppression of scientific data, the only published conclusion resulting from these experiments was that fluoride was safe in low doses, a profoundly helpful verdict for the US Military who feared lawsuits for fluoride injury from workers in nuclear power plants and munitions factories. Fluoride pollution was one of the biggest legal worries facing key US industrial sectors during the cold war. A secret group of corporate attorneys, known as the Fluorine Lawyers Committee, whose members included US Steel, ALCOA, Kaiser Aluminum, and Reynolds Metals, commissioned research at the Kettering Laboratory at the University of Cincinnati to "provide ammunition"(Bryson 2004) for those corporations who were fighting a wave of citizen claims for fluoride injury. The Fluorine Lawyers Committee and their medical ambassadors were in personal and frequent contact with the senior officials of the federal National Institute for Dental Research, and have been implied in the 'burying' of the forty year old Kettering study, which showed that fluoride poisoned the lungs and lymph nodes in laboratory animals. Private interests, sought to destroy careers and censor information by ensuring that scientific studies raising doubts about the safety of fluoride never got funded, and if they did, never got published.

During the 1990's, research conducted by Harvard toxicologist Phillis Mullenix showed that fluoride in water may lead to lower IQ's, and produced symptoms in rats strongly resembling attention deficit and hyperactivity disorder (ADHD). Just days before her research was accepted for publication, Mullenix was fired as the head of toxicology at the Forsyth Dental Center in Boston. Then her application for a grant to continue her fluoride and central nervous system research was turned down by the US National Institute of Health (NIH), when an NIH panel told her that "fluoride does not have central nervous system effects"(Griffiths 1998).

Despite growing evidence that it is harmful to public health, US federal and state pubic health agencies and large dental and medical organizations such as the American Dental Association (ADA), continue to promote fluoride. Water fluoridation continues, despite the Environmental Protection Agency (EPA)'s own scientists, whose union, Chapter 280 of the National Treasury Employees Union, has taken a strong stand against it. Dr. William Hirzy, vice president of Chapter 280, stated that "fluoride (that is added to municipal water) is a hazardous waste product for which there is substantial evidence of adverse health effects and, contrary to public perception, virtually no evidence of significant benefits"(Mullenix 1998). Although fluoride is up to fifty times more toxic than sulfur dioxide, it is still not regulated as an air pollutant by the American Clean Air Act. Since thousands of tonnes of industrial fluoride waste is poured into drinking water supplies all over North America, supposedly to encourage gleaming smiles in our children, big industry in the US has the benefit of emitting as much fluoride waste into the environment as they like with absolutely no requirement to measure emissions and no way of being held accountable for poisoning people, animals and vegetation.

In August 2003, the EPA requested that the National Research Council, the research arm of the National Academy of Sciences (NAS), re-evaluate water fluoride safety standards by reviewing recent scientific literature, because the last review in 1993 had major gaps in research. "Neither the US Food and Drug

Administration (FDA), nor the National Institute for Dental Research (NIDR), nor the American Academy of Pediatric Dentistry has any proof on fluoride's safety or effectiveness"(Sterling 1993). The International Academy of Oral Medicine and Toxicology has classified fluoride as an unapproved dental medicament due to it's high toxicity and the US National Cancer Institute Toxicological Program has found fluoride to be an "equivocal carcinogen" (Maurer 1990).

Currently the US government is continuing to introduce further fluoridation schemes throughout the country, including the Water Act passed in November 2003, which has made it impossible for water companies to undergo civil or criminal hearings as a result of adding fluoride to public water supplies.

In a society where products containing asbestos, lead, beryllium and many other carcinogens have been recalled from the marketplace, it is surprising that fluoride is embraced so thoroughly and blindly. It seems absurd that we would consider paying the chemical industry to dispose of their toxic waste by adding it to our water supply. Hiding the hazards of fluoride pollution from the public is a capitalist-style con job of epic proportions that has occurred because a powerful lobby wishes to manipulate public opinion in order to protect it's own financial interests. "Those who manipulate this unseen mechanism of society constitute an invisible government which is the true ruling power of our country... our minds are molded, our tastes formed, our ideas suggested, largely by men we have never heard of" (Bernays 1991).

Related Video on Edward Bernays and Mass Manipulation:
The Century of the Self: The Untold History Of Controlling The Masses Through The Manipulation Of Unconscious Desires

By Madison Ruppert Editor of <u>End the Lie</u>

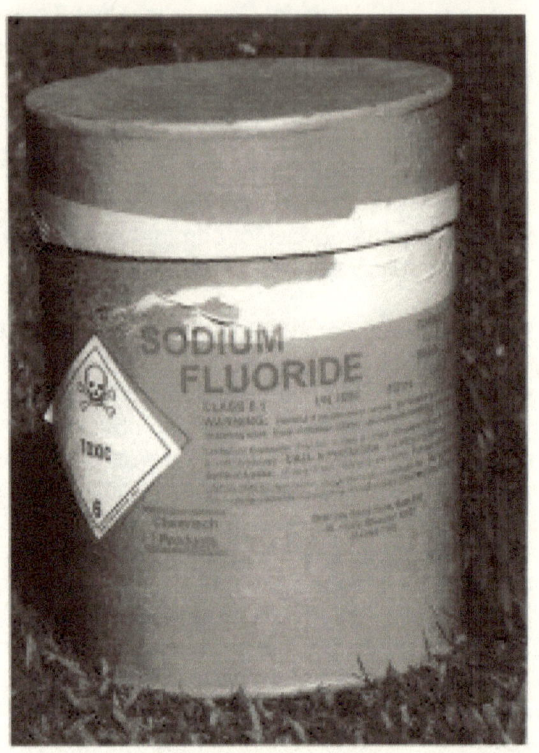

Despite a history of the criticism of water fluoridation being <u>characterized</u> as **"scare mongering via crazy water fluoridation conspiracy theories,"** <u>a study</u> conducted by the Harvard School of Public Health (HSPH) and China Medical University in Shenyang, combining 27 other studies, "found strong indications that fluoride may adversely affect cognitive development in children."

Hilariously, some actually go as far as to claim, "almost all health and dental organizations support the fluoridation of water or have *found no association with adverse effects with water fluoridation,"* which is <u>clearly untrue</u>.

The study, which was actually a systematic review of other studies, was **published on July 20, 2012 in Environmental Health Perspectives and mostly relied on research from China.**
The researchers focused on fluoride's interaction with the developing human brain since public health experts have so far been unable to come to a consensus on the safety of fluoridated drinking water consumption in children.

However, it has been established that "extremely high levels" of fluoride cause neurotoxicity in adults. Furthermore, negative impacts on learning and memory have been found in rodent studies, much like <u>high-fructose corn syrup</u>.

While proponents of water fluoridation regularly like to claim that there is plenty of science backing up the practice, according to Anna Choi, lead author of the study and research scientist in the Department of Environmental Health at HSPH, "Virtually no human studies in this field have been conducted in the U.S."

Therefore, the researchers relied on Chinese studies since, according to HSPH's <u>press release</u>, "exposures to the chemical are increased in some parts of China," and the risks of fluoride have been well documented there. Yet the researchers had to contend with the fact that some of the studies on children in China differed in various ways or were incomplete. This forced the authors to take care in data compilation and analysis in order to properly assess the potential risk of fluoride use.

"For the first time we have been able to do a comprehensive meta-analysis that has the potential for helping us plan better studies," said Choi. "We want to make sure that cognitive development is considered as a possible target for fluoride toxicity."

While the studies might have had some difficulties, the conclusions drawn by Choi and senior author and adjunct professor of environmental health at HSPH, Philippe Grandjean, seem to be quite clear.

Grandjean and Choi collated the various epidemiological studies of children who had been exposed to fluoridated drinking water and the China National Knowledge Infrastructure was also leveraged to find studies published in Chinese journals.

The authors then analyzed the relationship between fluoride exposure and IQ measures in over 8,000 children age 14 and below. **They discovered that in every study but one, highly fluoridated drinking water was linked to problems in cognitive development.**

That being said, the average loss was not as significant as some might expect. Averaged out, the loss was only half of a single IQ point, yet some studies indicated that even a slight increase in fluoride exposure could indeed be toxic to the brain.

The researchers further observed that children raised in areas with high fluoride content did, in fact, have significantly lower IQ scores than children who lived in areas with low fluoride.

The authors believe that the toxic effects of fluoride on brain development may have occurred long before the time the children were studied and that the brain may not be able to fully compensate for the toxic effects.

"Fluoride seems to fit in with lead, mercury, and other poisons that cause chemical brain drain," Grandjean says. "The effect of each toxicant may seem small, but the combined damage on a population scale can be serious, especially because the brain power of the next generation is crucial to all of us."

Personally, I find these statements to be nothing short of astounding. Usually we see fluoride being supported and touted, especially when discussing water fluoridation. These researchers, however, are presenting a much different picture, highlighting the fact that the cumulative damage of all of these toxicants is much more serious than some may want us to believe.

Did I forget anything or miss any errors? Would you like to make me aware of a story or subject to cover? Or perhaps you want to bring your writing to a wider audience? Feel free to contact me at admin@EndtheLie.com with your concerns, tips, questions, original writings, insults or just about anything that may strike your fancy.

Please support our work and help us start to pay contributors by doing your shopping through our Amazon link or check out some must-have products at our store.

Top Search Terms Used to Find This Page:

- *flouride*
- *Fluoride Containers*
- *fluoride is toxic*
- *fluoride poisoning*
- *fluoride toxic waste*

Help Spread Alternative News

*More at EndtheLie.com -
http://EndtheLie.com/2012/08/03/harvard-study-evidence-shows-fluoride-may-adversely-affect-cognitive-development-in-children/#ixzz2b3qYmxnz*

WAKE UP !!!
They are killing us !!

NOW FOR SOME MORE NASTY SURPRISES

YOU HAVE HEARD OF ALLERGY TO CERTAIN FOODS? MAYBE MILK, EGGS, ETC.?

I WAS ALWAYS AMAZED ABOUT AN ALLERGY TO PEANUTS. EATEN FOR THOUSANDS OF YEARS WITH IMPUNITY. NOW THEY CAN BE FATAL. WHY?

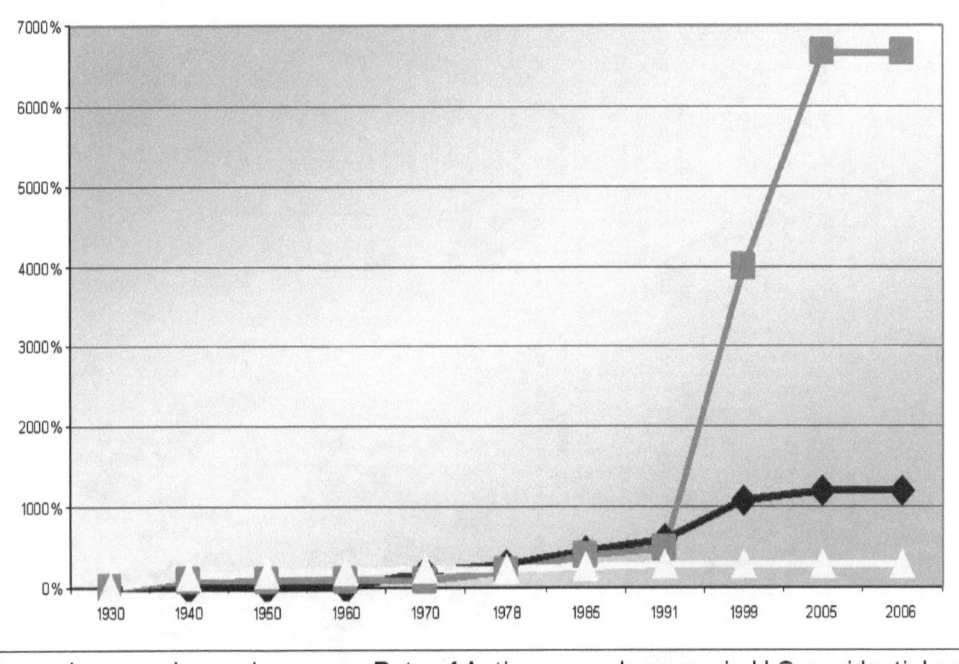

**THIS OFFICIAL GRAPH SHOWS THE INCEASE
IN AUTISM 1930-2006. THE 'SPIKE' IS OBVIOUS,
AND SYNCHRONOMOUS WITH VACCINES.**

**YET "SCIENTISTS" – "ARE BAFFLED BY RISE
IN AUTISM" AS IN LOCAL PRESS.** Press release
below.. Kind of reminds me of a previous statement known
globally and in my previous books (with amusement) "looks
like a duck, sounds like a duck, ...just like a duck. I wonder
what can it be?" (I leave the "duh" to you.)

Scientists baffled by rise in autism

b/ow Dom Post 28/11/11 LA15

Bob Brockie

WORLD OF SCIENCE

W HEN geeks breed they often produce more geeky children. So thinks Simon Baron-Cohen, professor of psychology at Cambridge University. The professor also worries that geeky couples may have more than their share of autistic offspring. He discovered this after studying a cluster of autistic children born to Silicon Valley parents and another batch at the Dutch technology hub town of Eindhoven.

Some clinicians back the professor's claims. A San Francisco psychologist says: "A lot of geeks don't make great eye contact, all their clothing comes from thrift shops and they don't have a lot of social understanding. When these geeks marry each other, that's bad news for the offspring."

Most autistic children are a sad problem. Many can't speak, few find work, daily activities are difficult and large numbers depend on their parents into adulthood. But Baron-Cohen finds that even these people may have extraordinary powers. Using a visual acuity and contrast test, the professor tested the sight of 15 autistic people in England and found they all had extremely sharp 20:7, hawk-like vision. Detailed facial features you and I can see

seven metres away were visible to these autistic people 20m away. They could read car number plates at the end of the street.

At the other end of the autism spectrum are the so-called idiot savants who can instantly multiply six figures by six figures in their heads, know the value of pi to 10,000 places or, like Stephen Wiltshire, can draw perfectly detailed cityscapes from memory after flying over them in a helicopter. Paradoxically, some of these people can't tie their own shoe laces.

But somewhere along the autism spectrum disorder are people whose

autism can be to their advantage. They are adept at spotting recurring patterns in large sets of data and don't forget things.

Danish entrepreneur Dr Thorkil Sonne runs an agency that has hired out more than 170 autistic people to perform nerdish work. He says that his autistic geeks stay focused beyond the point when most minds go numb and make for fewer mistakes.

"They have a preternatural capacity for concentration, near-total recall and strong attention to detail, are persistent and good at following structures and routines," he says. "In other words, they're born software engineers."

Since 2000, Dr Sonne's autists have taken jobs with companies such as Microsoft or Cisco ferreting out software errors or keeping track of fibre-optic networks.

Inspired by Dr Sonne, similar agencies have sprung up in Sweden, the Netherlands and the United States.

The number of autistic children has skyrocketed in recent years. The US Centre for Disease Control reports the incidence has jumped from 1:5000 children in 1975 to 1:110 children today. New Zealand has about 40 000 people "on the spectrum".

Older parenting accounts for some of the rise. Changing diagnostic criteria and greater awareness of the condition account for more of the increase but nearly 50 per cent of the rise remains unexplained. A review in this month's *Nature* science magazine admits that epidemiologists remain baffled.

Vaccines and the Peanut Allergy Epidemic

Tim O'Shea

Have you ever wondered why so many kids these days are allergic to peanuts? Where did this allergy come from all of a sudden?

Before 1900, reactions to peanuts were unheard of. Today almost **a 1.5 million children** in this country are allergic to peanuts.

What happened? Why is everybody buying EpiPens now?
Looking at all the problems with vaccines during the past decade, [2] just a superficial awareness is enough to raise the suspicion that vaccines might have some role in the appearance of any novel allergy among children.

But reactions to peanuts are not just another allergy. Peanut allergy has suddenly emerged as the **#1 cause of death from food reactions**, being in a category of allergens able to cause **anaphylaxis**. This condition brings the risk of asthma attack, shock, respiratory failure, and even death. Primarily among children.

Sources cited in Heather Fraser's 2011 book *The Peanut Allergy Epidemic* suggest a vaccine connection much more specifically. We learn that a class of vaccine adjuvants – **excipients** – is a likely suspect in what may accurately be termed an epidemic. [1]
But let's back up a little. We have to look at both vaccines and antibiotics in recent history, and the physical changes the ingredients in these brand new medicines introduced into the blood of children.

ANAPHYLACTIC SHOCK AND ALLERGY

Before 1900, anaphylactic shock was virtually unknown. The syndrome of sudden fainting, respiratory distress, convulsions, and sometimes death did not exist until vaccinators switched from the lancet to the hypodermic needle. That transformation was essentially complete by the turn of the century in the western world.

Right at that time, a new disease called **Serum Sickness** began to afflict thousands of children. A variety of symptoms, including shock, fainting, and sometimes death, could suddenly result following an injection.

Instead of covering it up, the connection was well recognized and documented in the medical literature of the day. **Dr Clemens Von Pirquet**, who actually coined the word "allergy," was a leading researcher in characterizing the new disease. [5] Serum Sickness was the first mass allergenic phenomenon in history. What had been required for its onset, apparently, was the advent of the hypodermic needle.

When the needle replaced the lancet in the late 1800s, Serum Sickness soon became a frequent visitor to the child's bed. It was a known consequence of vaccinations. Indeed, the entire field of modern allergy has evolved from the early study of Serum Sickness coming from vaccines.

VACCINE HYPERSENSITIVITY

Von Pirquet recognized that vaccines had 2 primary effects: **immunity** and **hypersensitivity**. [5] He said they were inseparable: the one was the price of the other.

In other words, if we were going to benefit from the effects of mass immunization, we must accept the downside of mass hypersensitivity as a necessary co-feature. Modern medicine has decided that this double effect should be kept secret, so they don't allow it to be brought up much.

Many doctors in the early 1900s were dead set against vaccines for this precise reason. The advertised benefit was not proven to be worth the risk. Doctors like Walter Hadwen MD, Wm. Howard Hay, and Alfred Russell Wallace saw how smallpox vaccines had actually increased the incidence of smallpox. [2,3] Wallace was one of the principal epidemiologists of the age, and his charts showing the increase in smallpox death from vaccination are unassailable – meticulous primary sources.

Another landmark researcher of the early 1900s was Dr Charles Richet, the one who coined the term **anaphylaxis**. [4] Richet focused on the reactions that some people seemed to have to certain foods. He found that with food allergies, the reaction came on as the result of **intact proteins** in the food having bypassed the digestive system and making their way intact into the blood, via leaky gut.

Foreign protein in the blood, of course, is a universal trigger for allergic reaction, not just in man but in all animals. [6]

But Richet noted that in the severe cases, food anaphylaxis did not happen just by eating a food. That would simply be food poisoning. Food anaphylaxis is altogether different. This sudden, violent reaction requires an initial **sensitization** involving injection of some sort, followed by a later ingestion of the sensitized food. Get the shot, then later eat the food.

The initial exposure creates the hypersensitivity. The second exposure would be the violent, perhaps fatal, physical event.

Richet's early work around 1900 was primarily with eggs, meat, milk and diphtheria proteins. Not peanuts. The value of Richet's research with reactive foods was to teach us the sequence of allergic sensitivity leading to anaphylaxis, how that had to take place.

Soon other doctors began to notice striking similarities between food reactions and the serum sickness that was associated with vaccines. Same exact clinical presentation.

PENICILLIN

Next up was penicillin, which became popular in the 1940s. It was soon found that additives called **excipients** were necessary to prolong the effect of the antibiotic injected into the body. The excipients would act as **carrier molecules.** Without excipients, the penicillin would only last about 2 hours. Refined oils worked best, acting as time-release capsules for the antibiotic.

Peanut oil became the favorite, because it worked well, and was available and inexpensive.

Allergy to penicillin became common, and was immediately recognized as a sensitivity to the excipient oils. To the present day, that's why they always ask if you're allergic to penicillin. The allergy is a sensitivity to the excipients.

By 1953 as many as 12% of the population was allergic to penicillin. [1] But considering the upside with life-threatening bacterial infections, it was still a good deal – a worthwhile risk.

By 1950 antibiotics were being given out like M&Ms. Soldiers, children, anybody with any illness, not just bacterial. Despite Alexander Fleming's severe warnings against prophylactic antibiotics, antibiotics were given indiscriminately as the new wonder drug. Just in case anything. [7] Only then, in the 1950s, did peanut allergy begin to occur, even though Americans had been eating peanuts for well over a century.

Remember – just eating peanuts cannot cause peanut allergy. Except if they are allowed to become moldy of course, in which case aflatoxins are released. But that's really not a peanut allergy.

When peanut allergy did appear, the numbers of cases were fairly small and initially it wasn't even considered worthy of study.

THE RISE OF VACCINES

The big change came with **vaccines**. Peanut oils were introduced as vaccine excipients in the mid 1960s. An article appeared in the NY Times on 18 Sept, 1964 that would never be printed today. [8] The author described how a newly patented ingredient containing peanut oil was added as an adjuvant to a new flu vaccine, in order to prolong the "immunity." The oil was reported to act as a time release capsule, and theoretically enhanced the vaccine's strength. Same mechanism as with penicillin.

That new excipient, though not approved in the US, became the model for subsequent vaccines. ([1] p 103)

By 1980 peanut oil had become the preferred excipient in vaccines, even though the dangers were well documented. [9] It was considered an **adjuvant** – a substance able to increase reactivity to the vaccine. This reinforced the **Adjuvant Myth**: the illusion that immune response is the same as immunity [2].

The pretense here is that the stronger the allergic response to the vaccine, the greater will be the immunity that is conferred. This fundamental error is consistent throughout vaccine literature of the past century.

Historically, researchers who challenged this Commandment of vaccine mythology did not advance their careers.

KEEPING PEANUT ADJUVANTS A SECRET

The first study of peanut allergies was not undertaken until 1973. It was a study of peanut excipients in vaccines. Soon afterwards, and as a result of the attention from that study, manufacturers were no longer required to disclose all the ingredients in vaccines. What is listed in the *Physicians Desk Reference* in each vaccine section is not the full formula. Same with the inserts. Suddenly after 1973, that detailed information was proprietary: the manufacturers knew it must be protected. Intellectual property. So now they only were required to describe the formula in general.

Why was peanut allergy so violent? Adjuvant pioneer Maurice Hilleman claimed peanut oil adjuvants had all protein removed by refining. [9] The FDA disagreed. They said some peanut protein traces would always persist [10]- that even the most refined peanut oils still contained some traces of intact peanut proteins. This was the reason doctors were directed to inject vaccines intramuscular rather than intravenous – a greater chance of absorption of intact proteins, less chance of reaction.

But all their secret research obviously wasn't enough to prevent sensitivity. Mother Nature bats last: no intact proteins in the body. 60 million years of Natural Selection didn't create the mammalian immune system for nothing. Put intact proteins, peanut or whatever, for any imagined reason into the human system and the inflammatory response will fire. And since the goal of oil emulsion adjuvants was to prolong reactivity in the first place – the notion of time-release – this led to sensitization.

PEANUT ALLERGY EPIDEMIC

Although peanut allergies became fairly common during the 1980s, it wasn't until the early 1990s when there was a sudden surge of children reacting to peanuts – the true epidemic appeared. What changed? The Mandated Schedule of vaccines for children doubled from the 80s to the 90s:

1980 – 20 vaccines
1995 – 40 vaccines
2011 – 68 vaccines

It would be imprudent enough to feed peanuts to a newborn, since the digestive system is largely unformed. But this is much worse – injecting intact proteins directly into the infant's body. In 36 vaccines before the age of 18 months.

A new kind of anaphylaxis appeared with peanut reactions: **reverse anaphylaxis.** (p 172, [1]) The reaction was not only to the sensitizing antigen, but to the weird new antibodies that had just been introduced in the human species by the new antigen. Without the usual benefit of the evolutionary process.

As vaccines doubled between the 1980s and the 1990s, hundreds of thousands of kids were now exhibiting peanut sensitivities, with frequent cases of anaphylaxis reactions, sometimes fatal.

But nobody talked about it.

Following the next enormous increase in vaccines on the Mandated Schedule after 9/11, whereby the total shot up to 68 recommended vaccines, the peanut allergy soon reached epidemic proportions: **a million children: 1.5% of them**. These numbers fit the true definition of epidemic even though that word has never been used in mainstream literature with respect to peanut allergy, except in Fraser's odd little book.

Many researchers, not just Heather Fraser, could see very clearly that

> **"The peanut allergy epidemic in children was precipitated by childhood injections."**
> **([1], p 106)**

But with the newfound research, the medical profession will do what they always must do – bury it. Protect the companies. So no money will be ever allocated from NIH to study the obvious connection between vaccine excipients and peanut allergy. That cannot happen, primarily because it would require a control group – an **unvaccinated population**. And that is the Unspoken Forbidden.

Same line of reasoning that has prevented Wakefield's work from ever being replicated in a mainstream US clinical study. No unvaccinated populations. Which actually means no studies whose outcome could possibly implicate vaccines as a source of disease or immunosuppression. Vaccines as a cause of an allergy epidemic? Impossible. Let's definitely not study it.

Instead let's spend the next 20 years looking for the Genetic Link to the childhood peanut allergy epidemic...

In such a flawed system, any pretense of true clinical science is revealed as fatally handicapped of course: we are looking for the truth, wherever our studies shall take us, except for this, and this, and oh yes, this.

Evidence for the connection between peanut excipients and vaccines is largely indirect today, because of the circling of the wagons by the manufacturers. It is very difficult to find peanut excipients listed in the inserts and PDR listings of vaccines. Simple liability.

FRAME OF REFERENCE

So in addition to all the other problems with vaccines delineated in our text, now we have a new one – peanut oil excipients. Which all by themselves can cause severe, even fatal, episodes of shock, as well as chronic allergy – irrespective of the mercury, aluminum, formaldehyde, ethylene glycol, and the attenuated pathogens which the manufacturers do admit to.

Quite a toxic burden to saddle the unprotected newborn with. No wonder the US Supreme Court refers to vaccines as "unavoidably unsafe."

Childhood allergies doubled between 1980 and 2000, and have doubled again since that time. [11] Theories abound. Childhood vaccines doubled at the same time. Why Is there a virtual blackout of viable discussion about this glaring fact?

The epidemic of peanut allergy is just one facet of this much broader social phenomenon. We have the sickest, most allergic kids of any country, industrialized or not, on Earth. A study of the standard literature of vaccines is identical to a study of the history of adjuvants – an exercise in cover-up and dissimulation.

Unvaccinated children don't become autistic. And they don't go into shock from eating peanuts.

But there can never be a formal clinical study where the control group is unvaccinated. NIH would never do that. They cannot. They know the outcome.

references

1. Fraser, H, *The Peanut allergy epidemic*, Skyhorse 2011
2. O'Shea, T, *Vaccination is not immunization*, thedoctorwithin 2013
3. Wallace, AR, *Vaccine delusion*, 1898
4. Richet, C, Nobel lecture, acceptance speech, 11 Dec 1913 Nobel Lectures Physiology or Medicine 1901-1921, Elsevier Publishing Company, Amsterdam, 1967
www.nobelprize.org/nobel_prizes/medicine/laureates/1913/richet-lecture.html
5. Von Pirquet, C, MD, On the theory of infectious disease *Journal of the Royal Society of Medicine* Volume 80, January 1987
6. O'Shea, T, Allergies: the threshold of reactivity
www.thedoctorwithin.com/allergies/Allergies-The-Threshold-of-Reactivity/
7. O'Shea, T, The post antibiotic age
www.thedoctorwithin.com/antibiotics/Post-Antibiotic-Age/
8. Jones, S, Peanut oil used in a new vaccine *New York Times* 18 Sep 13
9. HOBSON, D, MD, The potential role of immunological adjuvants in influenza vaccines *Postgraduate Medical Journal* March 1973 , no. 49, p 180.
http://pmj.bmj.com/content/49/569/180.full.pdf
9. Technical Report # 595, Immunological Adjuvants, World Health Org. 1976.
http://whqlibdoc.who.int/trs/WHO_TRS_595.pdf
10. FDA: March 2006. Approaches to Establish Thresholds for Major Food Allergens
www.fda.gov/downloads/food/labelingnutrition/foodallergenslabeling/guidancecomplianceregulatoryinformation/ucm192048.pdf

11. O'Shea, T, The threshold of reactivity
www.thedoctorwithin.com/allergies/Allergies-The-Threshold-of-
Reactivity/

**WE CAN CLEARLY SEE THE RESEARCH HAS
BEEN DONE, AND THE EVIDENCES ARE
BEFORE US. THE CAUSES AND EFFECTS ARE
UNDENIABLE. YET IT IS STILL CLAIMED
THAT "THERE IS NO SCIENTIFIC PROOF" "NO
SCIENTIFIC EVIDENCE".**

Then obviously there must be a defect with science. Or is
there? Could it be that "science" has prostituted itself?

I for one, must believe this is so.

Lets look at some past history of "science gone wrong"
in the names of vested interests.

<u>MEANWHILE BETWEEN BREAKS HERE ARE SOME INTERESTING AND OBVIOUS 'SCIENTIFIC 'COCK UPS'.</u>

The great expectations held for DDT have been realized. During 1946, exhaustive scientific tests have shown that, when properly used, DDT kills a host of destructive insect pests, and is a benefactor of all humanity.

Pennsalt produces DDT and its products in all standard forms and is now one of the country's largest producers of this amazing insecticide. Today, everyone can enjoy added comfort, health and safety through the insect-killing powers of Pennsalt DDT products . . . and DDT is only one of Pennsalt's many chemical products which benefit industry, farm and home.

GOOD FOR FRUITS—Bigger apples, juicier fruits that are free from unsightly worms . . . all benefits resulting from DDT dusts and sprays.

GOOD FOR STEERS—Beef grows meatier nowadays . . . for it's a scientific fact that—compared to untreated cattle—beef-steers gain up to 50 pounds extra when protected from horn flies and many other pests with DDT insecticides.

Knox-Out FOR THE HOME—helps to make healthier, more comfortable homes . . . protects your family from dangerous insect pests. Use Knox-Out DDT Powders and Sprays as directed . . . then watch the bugs "bite the dust"!

Knox-Out FOR DAIRIES—Up to 20% more milk . . . more butter . . . more cheese . . . tests prove greater milk production when dairy cows are protected from the annoyance of many insects with DDT insecticides like Knox-Out Stock and Barn Spray.

PENN SALT CHEMICALS

97 Years' Service to Industry • Farm • Home

GOOD FOR ROW CROPS—25 more barrels of potatoes per acre . . . actual DDT tests have shown crop increases like this! DDT dusts and sprays help truck farmers pass these gains along to you.

Knox-Out FOR INDUSTRY—Food processing plants, laundries, dry cleaning plants, hotels . . . dozens of industries gain effective bug control, more pleasant work conditions with Pennsalt DDT products.

"DDT is good for me-e-e!"

The Swiss chemist Paul Hermann Müller was awarded the Nobel Prize in Physiology or Medicine in 1948 "for his discovery of the high efficiency of dichlorodiphenyltrichloroethane as a contact poison against several arthropods."

In India thanks to DDT 0 people died in 1965, as in 1948 number was 3 million.
Greece, in 1938, had over one million people sick with malaria, as in 1959 this number was only 1200.
In February 1944, with a help of DDT, typhus epidemic was prevented in Naples.

But it turned out that DDT inevitably gets into the food chain. And on each link of the food chain there is an 10 times increase in the content of toxic DDT.

HOWEVER, ADDITIONAL TO LEAD PAINT, ASBESTOS AND OTHER HUMAN DISASTERS,

"SCIENCE" APPROVED OTHER PROBLEMS WERE TO BESET US,

(ARE YOU NOW BEGINNING TO ASK "IF YOU CAN TRUST 'SCIENCE'?")

The copy-only advertisement appeared in leading newspapers and magazines from March to June 1985 (Figure 1).

On June 16, 1986, the FTC issued a complaint against RJR, alleging that "Of cigarettes and science" falsely and misleadingly represents,

that the purpose of the MR FIT study was to determine whether heart disease is caused by cigarette smoking; that the MR FIT study provides credible scientific evidence that smoking is not as hazardous as the public or the reader has been led to believe; and that the MR FIT study tends to refute the theory that smok[...]

Figure 1. "Of cigarettes and science."

Of cigarettes and science.

This is the way science is supposed to work.

A scientist observes a certain set of facts. To explain these facts, the scientist comes up with a theory.

Then, to check the validity of the theory, the scientist performs an experiment. If the experiment yields positive results, and is duplicated by other scientists, then the theory is supported. If the experiment produces negative results, the theory is re-examined, modified or discarded.

But, to a scientist, both positive and negative results should be important. Because both produce valuable learning.

Now let's talk about cigarettes.

You probably know about research that links smoking to certain diseases. Coronary heart disease is one of them.

Much of this evidence consists of studies that show a statistical association between smoking and the disease.

But statistics themselves cannot explain why smoking and heart disease are associated. Thus, scientists have developed a theory: that heart disease is caused by smoking. Then they performed various experiments to check this theory.

We would like to tell you about one of the most important of these experiments.

A little-known study

It was called the Multiple Risk Factor Intervention Trial (MR FIT).

In the words of the *Wall Street Journal*, it was "one of the largest medical experiments ever attempted." Funded by the Federal government, it cost $115,000,000 and took 10 years, ending in 1982.

The subjects were over 12,000 men who were

thought to have a high risk of heart disease because of three risk factors that were statistically associated with this disease: smoking, high blood pressure and high cholesterol levels.

Half of the men received no special medical intervention. The other half received medical treatment that consistently reduced all three risk factors, compared with the first group.

It was assumed that the group with lower risk factors would, over time, suffer significantly fewer deaths from heart disease than the higher risk factor group.

But that is not the way it turned out.

After 10 years, there was no statistically significant difference between the two groups in the number of heart disease deaths.

The theory persists

We at R.J. Reynolds do not claim this study proves that smoking doesn't cause heart disease. But we do wish to make a point.

Despite the results of MR FIT and other experiments like it, many scientists have not abandoned or modified their original theory, or re-examined its assumptions.

They continue to believe these factors cause heart disease. But it is important to label their belief accurately. It is an opinion. A judgement. But *not* scientific fact.

We believe in science. That is why we continue to provide funding for independent research into smoking and health.

But we do not believe there should be one set of scientific principles for the whole world, and a different set for experiments involving cigarettes. Science is science. Proof is proof. That is why the controversy over smoking and health remains an open one.

R.J. Reynolds Tobacco Company

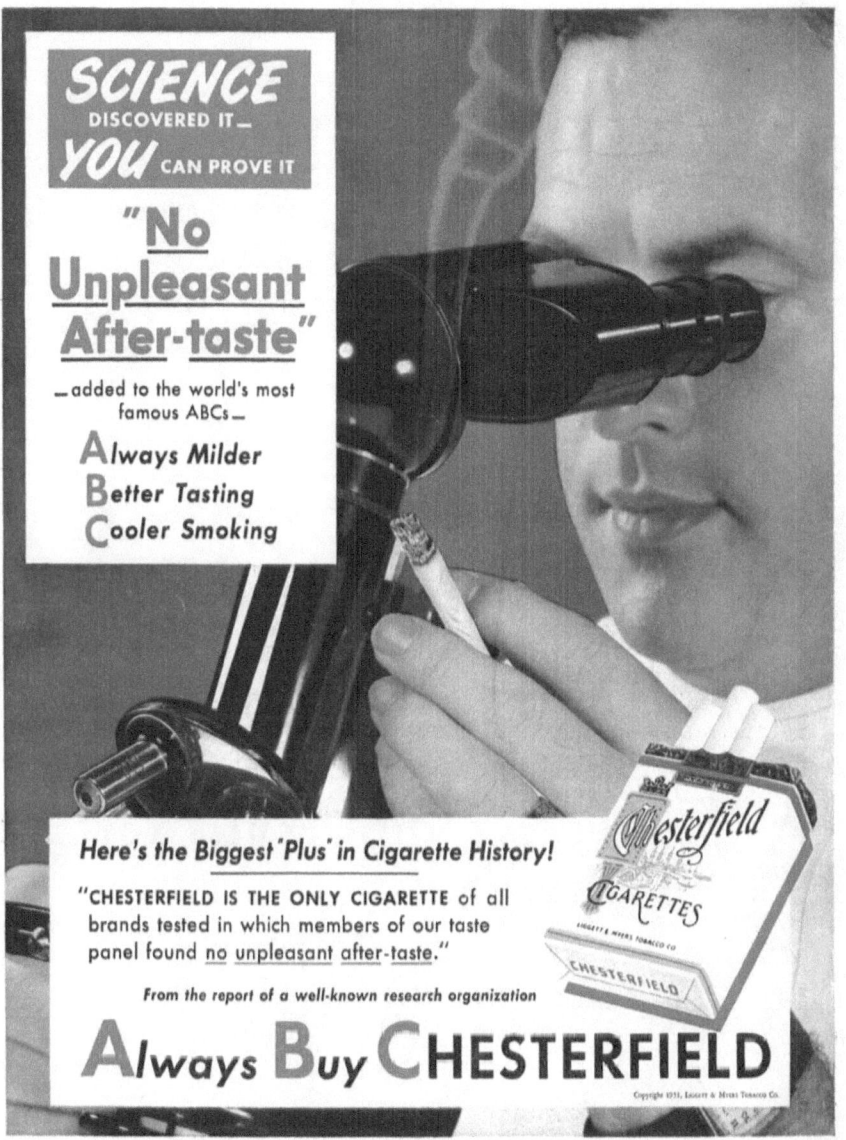

SCIENCE
DISCOVERED IT—
YOU CAN PROVE IT

"No Unpleasant After-taste"

—added to the world's most famous ABCs—

Always Milder
Better Tasting
Cooler Smoking

Here's the Biggest "Plus" in Cigarette History!

"CHESTERFIELD IS THE ONLY CIGARETTE of all brands tested in which members of our taste panel found no unpleasant after-taste."

From the report of a well-known research organization

Always Buy CHESTERFIELD

AH. THERE IT IS. THE TESTS SHOW. IS THAT
SCIENTIFIC PROOF? THEY THINK SO.

HERE IS APPROVAL.

He's one of the busiest men in town. While his door may say *Office Hours 2 to 4*, he's actually on call 24 hours a day.

The doctor is a scientist, a diplomat, and a friendly sympathetic human being all in one, no matter how long and hard his schedule.

According to a recent Nationwide survey:

MORE DOCTORS SMOKE CAMELS THAN ANY OTHER CIGARETTE

DOCTORS in every branch of medicine—113,597 in all—were queried in this nationwide study of cigarette preference. Three leading research organizations made the survey. The gist of the query was—What cigarette do you smoke, Doctor?

The brand named most was Camel!

The rich, full flavor and cool mildness of Camel's superb blend of costlier tobaccos seem to have the same appeal to the smoking tastes of doctors as to millions of other smokers. If you are a Camel smoker, this preference among doctors will hardly surprise you. If you're not—well, try Camels now.

Your "T-Zone" Will Tell You...

T for Taste . . .
T for Throat . . .

that's *your* proving ground for any cigarette. See if Camels don't suit *your* "T-Zone" to a "T."

R. J. Reynolds
Tobacco Company
Winston-Salem, N. C.

CAMELS *Costlier Tobaccos*

It obviously developed?

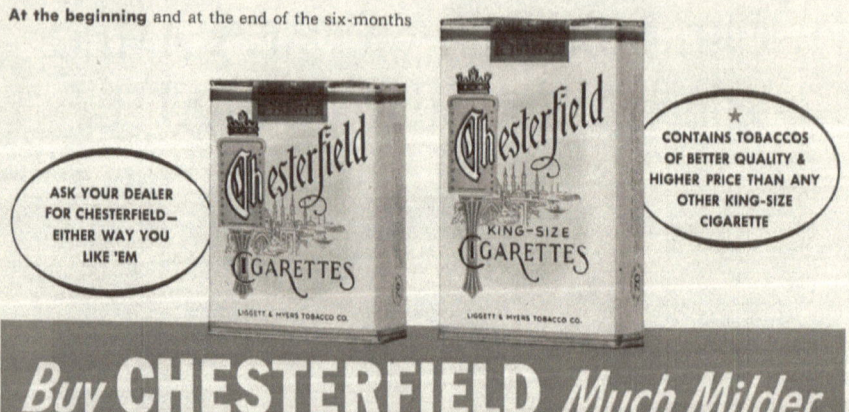

NOSE, THROAT,

and Accessory Organs not Adversely Affected by Smoking Chesterfields

FIRST SUCH REPORT EVER PUBLISHED ABOUT ANY CIGARETTE

A responsible consulting organization has reported the results of a continuing study by a competent medical specialist and his staff on the effects of smoking Chesterfield cigarettes.

A group of people from various walks of life was organized to smoke only Chesterfields. For six months this group of men and women smoked their normal amount of Chesterfields – 10 to 40 a day. 45% of the group have smoked Chesterfields continually from one to thirty years for an average of 10 years each.

At the beginning and at the end of the six-months period each smoker was given a thorough examination, including X-ray pictures, by the medical specialist and his assistants. The examination covered the sinuses as well as the nose, ears and throat.

The medical specialist, after a thorough examination of every member of the group, stated: "It is my opinion that the ears, nose, throat and accessory organs of all participating subjects examined by me were not adversely affected in the six-months period by smoking the cigarettes provided."

ASK YOUR DEALER FOR CHESTERFIELD— EITHER WAY YOU LIKE 'EM

CONTAINS TOBACCOS OF BETTER QUALITY & HIGHER PRICE THAN ANY OTHER KING-SIZE CIGARETTE

Buy CHESTERFIELD—*Much Milder*

What to most people is NOT obvious is that the medical (and Dental) profession DID SUPPORT THESE NEFARIOUS ADDICTIONS AND DRUGS, AND GAINED MONEY AND SUPPORT FROM THEM. But not only them. Consider.

Yeah, back then it was smoke "Kent" or "camel;" because they PAID. (Your 'student debt was easy to repay – if at all.) But what happened when the 'tobacco' industry money dried up? Where is a new "milk money"?

People and/or celebrities were used to support the 'cause'.

NOW...10 Months Scientific Evidence For Chesterfield

A MEDICAL SPECIALIST is making regular bi-monthly examinations of a group of people from various walks of life. 45 percent of this group have smoked Chesterfield for an average of over ten years.

After ten months, the medical specialist reports that he observed...

no adverse effects on the nose, throat and sinuses of the group from smoking Chesterfield.

MUCH MILDER

CHESTERFIELD IS BEST FOR YOU

Copyright 1953, Liggett & Myers Tobacco Co.

First and Only Premium Quality Cigarette in Both Regular and King-Size

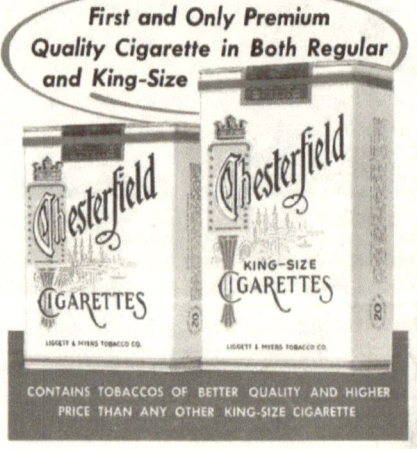

CONTAINS TOBACCOS OF BETTER QUALITY AND HIGHER PRICE THAN ANY OTHER KING-SIZE CIGARETTE

The famous Perry Como lived till about 89, and died in sleep it is reported. The Altziemers got there first.

Then we had and were faced with blatant (now) rubbish...

Seriously people. I have hundreds of illustrations that would embarrass you or me, the above are a mere collection from the above. The Internet reveals MILLIONS. What is evidenced? Be real. What? Did "science" become involved with a financial remuneration from such things as tobacco, mercury use, or indeed many other "dubious" medical or other practices? In those I have not mentioned 'fluoride', 'vaccines', or a hundred equal or more nefarious activities foisted upon mankind by the corporates in the name and pursuit of pure profit.

It cannot be in the interest of science or medicine for the benefit of mankind, were that so a lot of the trash would be removed from distribution and availability as soon as evidence of debilitating effects of their trash was revealed.

We must logically conclude that the purveyors and distributors of such 'garbage' have no interest other than profit, and also power and control.

We must also logically conclude that there is an active process of systematic denial or responsibility of or for any defect in products being presented as efficacious. The frequent use of phrases such as "there is no scientific proof or evidence....." clearly demonstrate absolute <u>denial</u> when the evidence is plainly before everyone who will simply look. Perhaps the class of person who says such things is worse than a mere 'errorist", but a willing stooge or active conspirator.

The following two clips from a local newspaper very nicely illustrate this active process of "denial" and claim there is no scientific proof or evidence. It is in relation to the fluoridation of the Hutt Valley in New Zealand.

First the clips, then the comments they generate.

Supporters under fire

...dation supporters have been ...ged to "get up to speed" during a ...bate at a meeting of the Hutt Val-... District Health Board.

The board adopted a paper on ...day endorsing the ongoing fluori-...tion of local water supplies.

Health board member Ken Laban ...d he had been impressed with ... anti-fluoride lobby when they ...ke at Hutt City Council.

"I don't think they're just a rad-ical group that you can just brush away."

Fellow board member and former Lower Hutt mayor David Ogden agreed that the anti-fluoridation movement should not be underestimated.

"My personal view is that people who are arguing our case really need to get up to speed."

Board member and Carterton mayor Ron Mark said the anti-fluoridation proponents were vocal and well-organised, but he had heard pretty much all of their arguments.

"Our greater fear is the amount of coca cola and soft drinks with high sugar content that children are drinking."

Hutt deputy mayor and board member David Bassett said aquifers were a good way of meeting the needs of people who did not want their water fluoridated.

Bassett said the taps in Petone with a direct supply to the under-ground water resource were popu-lar. The council had plans to build two further bores to the aquifer in Waterloo and Taita.

Board member John Terris said the health board paper was unbalanced because it did not pro-vide an accurate summary of the opponents of fluoridation.

However, Dr Ashleigh Bloomfield, who was present at the board meet-ing and endorsed the report, said the paper was centred to reflect the weight of scientific evidence.

► FLUORIDE DEBATE

SUPPORTERS SAY
■ Fluoride is proven to reduce tooth decay.
■ There are no proven health risks at levels used.
■ The 2009 Ministry of Health oral health survey suggested children, adolescents and adults living in fluoridated areas had less decay than those in non-fluoridated areas.
■ There was no significant difference in rates of dental fluorosis (marking of teeth from exposure to excessive fluoride) between fluoridated and non-fluoridated areas.

OPPONENTS SAY
■ It causes dental fluorosis.
■ It causes increased rates of cancer, premature birth, kidney problems, heart disease, low IQ, bone problems, thyroid damage and other health issues.
■ The small benefit to dental health does not outweigh the risk.

"I don't think they're just a radical group that you can just brush away."

Fellow board member and former Lower Hutt mayor David Ogden agreed that the anti-fluoridation movement should not be underestimated.

"My personal view is that people who are arguing our case really need to get up to speed."

Board member and Carterton mayor Ron Mark said the anti-fluoridation proponents were vocal and well-organised, but he had heard pretty much all of their arguments.

"Our greater fear is the amount of coca cola and soft drinks with high sugar content that children are drinking."

Hutt deputy mayor and board member David Bassett said aquifers were a good way of meeting the needs of people who did not want their water fluoridated.

Bassett said the taps in Petone with a direct supply to the underground water resource were popular. The council had plans to build two further bores to the aquifer in Waterloo and Taita.

Board member John Terris said the health board paper was unbalanced because it did not provide an accurate summary of the opponents of fluoridation.

However, Dr Ashleigh Bloomfield, who was present at the board meeting and endorsed the report, said the paper was centred to reflect the weight of scientific evidence.

I feel the first clip illustrates a prejudice in favour of the pro-fluoride errorists. The right hand side comments "fluoride debate" give specific details of the errorists, and then very scant if any information of "opponents say". Not a 'biggie, but significant bearing in mind this is the media reporting what should be an extremely important issue and 'debate'

Frankly I felt the report turns the "debate" actually into a "circus".

The information in paras 2 and 3 clearly indicate former mayor Ogden's support for pro-fluoride by advising they need to to get "up to speed" as the "anti-fluoridation" movement (that really is the educated public to which he dismisses as a 'movement.) should nor be underestimated. Note: not listened to but treated as a protagonist to be cautious of.

Para 4 says it all. Board member and also a mayor (implied authority so listen to him) "has heard pretty much all of their arguments…" This is tantamount to saying, "I treat this as an argument, I know what they say and think, I am not listening, not looking, my mind is closed and made up". Typical of denialist errorist stooges.

The Truth About Gardasil

March 28, 2013
By admin

Gardasil is the HPV vaccine, touted to fight cervical cancer. What they are not telling you is that thousands of girls are having adverse reactions to the HPV Vaccines, some have even died -at last count, at least 103 lives have been lost. We have got to do something about this. These girls need our help!

These girls are having reactions such as; seizures, strokes, dizziness, fatigue, weakness, headaches, stomach pains, vomitting, muscle pain and weakness, joint pain, auto-immune problems, chest pains, hair loss, appetite loss, personality changes, insomnia, hand/leg tremors, arm/leg weakness, shortness of breath, heart problems, paralysis, itching, rashes, swelling, aching muscles,pelvic pain, nerve pain, menstrual cycle changes, fainting, swollen lymph nodes, night sweats, nausea, temporary vision/hearing loss just to name some of them!

There is no known treatment to help these girls, as they suffer in silence. The doctors, if they even admit the connection, have no idea how to help them. So they spend their days going from appointment to appointment, from specialist to specialist trying to find someone to help them. Many of these families have started looking for help outside of mainstream medicine, which in some cases, may bring minor relief. However, most insurance plans do not cover this type of treatment, and as a result, this route is out of reach for many girls.

As you look at the faces on this page, remember that these are just some of the thousands of girls whose lives have been affected by Gardasil. Sadly, one of the girls pictured above, Jessie Ericzon, passed away just two days after her third gardasil injection.

If you or someone you know is experiencing any side effects or you would just like to contact us, you can do so at moderator@truthaboutgardasil.org or you can log into our Gardasil **forum** to tell your story.

Personal Accounts of the Gardasil Injured

There are so many people that have been injured by the gardasil vaccine, that we have decided to have a spotlight page. The injured have all been through so much, and deserve to have their stories told. The words you will read here are their own personal accounts. Just be prepared, for what you will read here is heartbreaking.

I am the mother of my precious little girl Alisa. Though she is not little anymore, she will always be my princess. Alisa grew up healthy, active, and happy. She enjoyed trying new things and being around others. She loved photography, fishing, bike riding, boating, hiking, martial arts, music (especially her violin), and all types of animals. In high school she enjoyed her photography in which she won an awards for. She was a tattoo assistant for a Washington State award winning tattoo artist and was learning the trade. She was on the swim team and loved swimming. Being a concerned mother I was fearful for her getting cervical cancer. We were being bombarded with commercials and ads for this vaccine for girls to be one less. When they offered the vaccine to my daughter I said sure. They never went over side effects or problems, so I figured it was a safe one. That day is when our nightmare began.

August 14, 2007, the first shot (lot # 0384U)- Alisa went home with her injection site itchy, swollen, red, and sore. She was not feeling well, like flu. I sent her to bed and she was home sick for a couple days. **The doctors office said that was normal**. I did notice she was complaining about sore joints and muscles in the months to follow and she was napping more with headaches. I just blamed it on growing pains.

November 14, 2007, 2nd shot (lot # 0927U)-**Alisa was not thrilled about having the shot again. She was so scared** she was going to feel crummy again. Sure enough the injection site was itchy, sore, red and this time the area swelled up a large area. She went home and went to bed. She was out of school again down with flu like symptoms for days. Slowly she came out of her slump but was so sore throughout her body. Sore muscles and joints, complained of headaches. She was exhausted all the time. It was difficult for her to muster the strength to do things. She took to her computer and was playing games online with friends.

Happy Valentines Day to you....February 14, 2008 (lot# 12APR10) **This time she was flat terrified and cried all the way to the doctors begging me not to make her get the shot.** This breaks my heart because I remember this day too vividly. **I kept telling her it was the last shot and we don't want to make the other two shots go without the final one. I kept telling her it was for the best.** We left the office and she was throwing up, shaking, feverish, and the usual symptoms of sore muscles and joints, a pounding headache, exhaustion. She was down for over a week recouping from the shot.

After the 3 shots I put the series out of my mind. The only time that **I remember the shots was EVERY visit to the hospitals and Dr.s offices, when they would ask if Alisa had all her vaccines. EVERY time I answered I said, "yes she has even had the Gardasil series." Not one doctor put this together.** The cause to her illnesses were put together by good friends of the family.

Alisa continued on with her life but lost some of her spunkiness. She no longer had tons of energy. She slept a bunch, stopped her swim team, and spent more time in her room on the computer. She said her body hurt and didn't feel like doing anything.

There are so many doctors appointments and hospital visits in this time frame. I have requested all her records from the hospitals, clinics, and her primary care provider (he has discharged her from his office).

October 2009, another visit to the emergency department with eye problems. She was having bloody discharge and pressure behind her right eye. Alisa was having problems with slurred speech, headache and facial droop.They account it to pink eye and we begin treatment. They recommend we contact a neurologist and see the doctor.

After seeing the neurologist finally we were sent home with the idea of further testing later. Later that night the neurologist called recommending we take her back to the hospital for further testing. **October 2009, off to the hospital again. By this time, Alisa was admitted into children's hospital with the following symptoms: Bells Palsy, migraine, right sided weakness, blurry vision, tinnitus, balance problems, numbness right side, unable to walk, problems swallowing, fatigue, joint pain, difficulty in opening mouth. They were testing for stroke and other unknown causes to this problem.** She endured CAT scans, MRI's, Lumbar punches. Over the next week she continued to get worse. Of course the psychiatric doctors were sent in to ensure she wasn't an abuse victim. We didn't know yet then she was abused by the pharmaceutical company. **She was discharged without a cause to the problem.**

This was Alisa's senior year in high school, though the first semester she was in the hospital and a tutor came in a few times a week to drop off and pick up homework. The doctors released her to return to school but the noises, medications, and lights caused her headaches to pound and she was struggling with anxiety issues. With only one semester until graduation she dropped out of school. It was too much to handle. (She tested and passed her GED in Dec. 2011).

Over the following years Alisa has had this happen 2 more times. Right side paralysis, wheelchair, learn to walk with a walker again, and now she gets so exhausted she uses the wheelchair to save her energy.

Her side effects seem to increase in intensity and keep adding in numbers. **So far she struggles with the following issues: leaky gut syndrome, pins and needles in extremities, dizziness, bleeding gums, toothaches/teeth changes, brain fog, sensitivity to chemicals, chest pains, constipation, dehydration, enlarged liver, sound sensitivity w/anxiety,**

extreme pain in the tailbone area, fainting, fever and blisters, fibromyalgia, <u>Guillain-Barre syndrome</u>, autistic-like symptoms, hand/leg weakness, back pain, hot//cold intolerance, trouble sleeping, itching, joint pain, knee pain, light sensitivity, blindness, depression, personality changes, anxiety/panic attacks, loss of bladder control, bladder issues, muscle aches and spasms, muscle tension, tumor, paleness, chronic fatigue syndrome, paralysis, pneumonia, severe nerve pain, shortness of breath, slurred speech, smell sensitivity, diarrhea, sore throat, stomach pain, swelling/edema, tremors hand and/or leg, random twitching of extremities, bloating, uterine spasms, hair loss, urinary tract infections (UTI), kidney issues, vision loss (temporary/permanent), vision problems, dyslexia, hallucinations, vomiting blood, stomach ache, nausea, rashes, appetite loss, weight gain or loss (20-30lbs).

Alisa is unable to anything without supervision. Showering requires a shower chair. She has to be checked on constantly since her seizures come without notice. She becomes so disoriented after some of the seizures, she wanders off.

Alisa was once firm in her beliefs about anti-smoking and drug use, but has now become a medical marijuana patient. This is the only thing that takes an edge off the pain. She is never pain free but the MMJ makes it more tolerable. **Unfortunately this pain relief isn't covered by medical insurance so the $1200.00 a month pain control comes out of the pocket book.**

This vaccine has dramatically changed our family's life. Alisa fights for her life everyday. It is even more difficult to not be able to help her or find a cure for her symptoms. It is heartbreaking to watch your child suffer in horrible pain and not be able to help. **I feel guilty because my child is suffering because of a choice I made. A choice I thought was to help her and instead has disabled her**. <u>I wish someone would have told me. Please help spread the word about this vaccine. Tell everyone you know about the dangers of the vaccine. Educate before you vaccinate. This vaccine is harming thousands of girls, and now boys. The vaccine needs to be taken off the market. The numbers of children affected is rising everyday.</u>

Hi my name is Nicole Marie Goodman. I am 18 years old .I believe that the gardasil is the reason for my current state. I could not believe that a vaccine could truly hurt and injure all these innocent girls.

I had no idea about the shot I just heard prevent cervical cancer and I was all for it but never once was I told that it could hurt, paralyze and for some severe reactions unfortunately cause death for and young girls and women. We have to put a stop to this and take the vaccine off the market completely. I would like to share my story with girls who have traveled the same road I have.

My journey begins on one night on I came home from cheerleading practice and I was exhausted and had later on that night. so I decided to take a quick cat nap to rest up for my fun night, I was with my friend and we were home alone at the time my mother and siblings were at the ymca. I started slowly waking up from my nap still have asleep. until I really woke up to the most excruciating pain in my right calf. I told my friend and she tried to see if she could message it an rub it out thinking it was a simple muscle cramp. so after she did that the pain just started excluding until I suddenly found myself in the worst pain with tears streaming down my face .so I turned to my friend and asked her to call my mother and tell her to come home. my mother came home and asked me what had happened so I told her so after telling her what had happened she suggest that I try icy hot which unfortunately made the pain worst.

So after an hour my mother and I decided we should go to the emergency room to get it checked out. we arrived at the e.r. and they put a shot in my muscle relax cent directly into my calf muscle using a syringe and told me to sit patiently for about an hour to see if it would respond to the muscle cramp. it did not so I was discharged and sent home with narcotics was told to follow up with my primary .so I went on two months living on narcotics because that was the only way I did not find myself in pain. so about a month went on feeling just fine until my right calf again to twitch so

I was sent to an orthopedic which then told me I could have a possible tumor in my head .so on April first I was admitted to children's hospital. Were I was cleared for my brain. I stayed in children for a week ran test and nothing acute showed up was diagnosed with kinetic movement disorder. Then I was sent to get an afo made from nopco. Which was around may when I was fitted for that and went to the nopco clinical. on april fourth I had my first episode .i felt perfectly fine that night I was cleaning my room and all I remember doing was getting up to go to the bathroom and apparently did not make it there. my mother heard the big bang and had my sister go see where it came from my sister go see where it came from my sister checked every room just when she thought she was done she remembered she forgot my room.

So she opened the door and there she found me past out on the floor my mother came running in and called 911 after calling the paramedics arrived in a matter of five minutes and transferred me to Tobey hospital in Wareham. There I came conscious but was still lethargic and still out of it. at Tobey they did multiple tests and blood work after those test were done .they deci8ded to transport me to mass general hospital. when I arrived at M .G.H. more tests were done and was admitted to the hospital for a week .i was then discharged to go home on crutches and was told to follow up with pt ant ot and see my primary doctor.

Then in October found myself back into the er with chronic abdomen pain and persistent vomiting and unexplained weight loss. then I was admitted for gi problems and they're impression was that I had an anomaly in small bowel series. they ran test again also such as mri ekg eeg emg iv fluids and lab work .when testing came back I had an abnormality on my ekg and was placed in the icu for my heart. all had a cope done to see how much acid was in my stomach along with checking for a possible ulcer .after all the gi tests I found myself right back in the er in both hospitals with more gi problems. the doctors could never actually find a cause for my gi problems nor my drastic weight lost. the symptoms just slowly started to drift away. then I was sent to pt and ot again to be fitted to a kfo because the tremor was progressing . and on my fifth I had another fainting episodes .i was perfectly fine driving back from night school I pulled in my drive way and all I remember was shutting the door to my jeep.

I blacked out completely and I walked up my stairs and there I collapsed again. Wasn't as bad as last episode in October though. so from my house I was transferred to Tobey to be checked out then transferred to MGH to be evaluated by the neurology team. where there I dealt with ignorant doctors telling me it was in my head. so I was sent home in the wheel chair not mobile. I was home extremely weak for a week and then a week later found myself hard as a rock with tone throughout my legs. weeks went on till I was sent to a physiatrist were he told me it is defiantly dystonia and begged me to give children's hospital a second chance ,more like a thousand. so I arrive at children's and got admitted and have more tests repeated and a baclofen pump trial. the trial was done and got positive results out of it. I was told I would be getting the surgery until two hours later when they came in specifically told me it worked and could pe a possible step to recovery. later that day my parents and I had a team meeting with the neurology team and they told me the reason why they put the trial aside because they were convinced it was conversion disorder.

So after hearing that it was in my head for the second time I demanded to be discharged that night. there I went home saw the physiatrist and he told me he would get that pump in me .so I recently just had another trial done and once again had positive outcomes from it. While my diagnosis was conversion disorder no one wanted to touch me with a ten foot pole. So my tone progressed and spread to my bladder and arm. Now I am cathing myself and wearing a sling on my left arm because of the tone and pain .That is my current were I stand right now

I would like to thank everyone that took the time to read my story to all the gardasil girls out there we will make it through this and find a cure. and for any parents or girls that would like to ask questions about my story or symptoms you can email me at nikkigood3@gmail.com or if you're a member of face book you can friend request me .
Thanks   again
Nikki (Nicole Goodman)

512-THE BITTER TRUTH ABOUT

ARTIFICIAL SWEETENERS

"A man may fish with the worm that hath eat of a king, and eat of
the fish that hath fed of that worm." - Shakespeare, Hamlet

Aspartame sugar substitutes cause worrying symptoms from memory loss
to brain tumours. But despite US FDA approval as a 'safe' food additive,
aspartame is one of the most dangerous substances ever to be foisted
upon an unsuspecting public.

Aspartame is the technical name for the brand names, NutraSweet, Equal,
Spoonful, and Equal-Measure. Aspartame was discovered by accident in
1965, when James Schlatter, a chemist of G.D. Searle Company was
testing an anti-ulcer drug. Aspartame was approved for dry goods in 1981
and for carbonated beverages in 1983. It was originally approved for dry
goods on July 26, 1974, but objections filed by neuroscience researcher
Dr John W. Olney and Consumer attorney James Turner in August 1974
as well as investigations of G.D. Searle's research practices caused the US
Food and Drug Administration (FDA) to put approval of aspartame on
hold (December 5, 1974). In 1985, Monsanto purchased G.D. Searle and
made Searle Pharmaceuticals and The NutraSweet Company separate
subsidiaries.

Aspartame is, by far, the most dangerous substance on the market that is
added to foods. **Aspartame accounts for over 75 percent of the adverse
reactions to food additives** reported to the US Food and Drug
Administration (FDA). Many of these reactions are very serious including
seizures and death as recently disclosed in a February 1994 Department
of Health and Human Services report.(1) A few of the 90 different
documented symptoms listed in the report as being caused by aspartame
include: Headaches/migraines, dizziness, seizures, nausea, numbness,
muscle spasms, weight gain, rashes, depression, fatigue, irritability,
tachycardia, insomnia, vision problems, hearing loss, heart palpitations,
breathing difficulties, anxiety attacks, slurred speech, loss of taste,
tinnitus, vertigo, memory loss, and joint pain.

According to researchers and physicians studying the adverse effects of aspartame, the following chronic illnesses can be triggered or worsened by ingesting of aspartame:(2) Brain tumors, multiple sclerosis, epilepsy, chronic fatigue syndrome, Parkinson's disease, Alzheimer's, mental retardation, lymphoma, birth defects, fibro myalgia, and diabetes.

Aspartame is made up of three chemicals: Aspartic acid, phenylalanine, and methanol. The book, Prescription for Nutritional Healing, by James and Phyllis Balch, lists aspartame under the category of "chemical poison." As you shall see, that is exactly what it is.

ASPARTIC ACID (40% OF ASPARTAME)

Dr Russell L. Blaylock, a professor of Neurosurgery at the Medical University of Mississippi, recently published a book thoroughly detailing the damage that is caused by the ingestion of excessive aspartic acid from aspartame. [Ninety nine percent of monosodium glutamate 9MSG) is glutamic acid. The damage it causes is also documented in Blaylock's book.] Blaylock makes use of almost 500 scientific references to show how excess free excitatory amino acids such as aspartic acid and glutamic acid in our food supply are causing serious chronic neurological disorders and a myriad of other acute symptoms.(3)

SUMMARY OF HOW ASPARTATE (AND GLUTAMATE) CAUSE DAMAGE

Aspartate and glutamate act as neurotransmitters in the brain by facilitating the transmission of information from neuron to neuron. Too much aspartate or glutamate in the brain kills certain neurons by allowing the influx of too much calcium into the cells. This influx triggers excessive amounts of free radicals which kill the cells. The neural cell damage that can be caused by excessive aspartate and glutamate is why they are referred to as "excitotoxins." They "excite" or stimulate the neural cells to death.

Aspartic acid is an amino acid. Taken in its free form (unbound to proteins) it significantly raises the blood plasma level of aspartate and glutamate. The excess aspartate and glutamate in the blood plasma shortly after ingesting aspartame or products with free glutamic acid (glutamate precursor) leads to a high level of those neurotransmitters in certain areas of the brain.

The blood brain barrier (BBB) which normally protects the brain from excess glutamate and aspartate as well as toxins 1) is not fully developed during childhood, 2) does not fully protect all areas of the brain, 3) is damaged by numerous chronic and acute conditions, and 4) allows seepage of excess glutamate and aspartate into the brain even when intact.

The excess glutamate and aspartate slowly begin to destroy neurons. The large majority (75%+) of neural cells in a particular area of the brain are killed before any clinical symptoms of a chronic illness are noticed. A few of the many chronic illnesses that have been shown to be contributed to by long-term exposure excitatory amino acid damage include:

Multiple sclerosis (MS), ALS, memory loss, hormonal problems, hearing loss, epilepsy, Alzheimer's disease, Parkinson's disease, hypoglycaemia, AIDS dementia, brain lesions, and neuroendocrine disorders.

The risk to infants, children, pregnant women, the elderly, and persons with certain chronic health problems from excitotoxins are great. Even the Federation of American Societies For Experimental Biology (FASEB), which usually understates problems and mimics the FDA party line, recently stated in a review that "it is prudent to avoid the use of dietary supplements of L-glutamic acid by pregnant women, infants, and children. The Existence of evidence of potential endocrine responses, i.e., elevated cortisol and prolactin, and differential responses between males and females, would also suggest a neuroendocrine link and that supplemental L-glutamic acid should be avoided by women of childbearing age and individuals with affective disorders."(4) Aspartic acid from aspartame has the same deleterious effects on the body as glutamic acid.

The exact mechanism of acute reactions to excess free glutamate and aspartate is currently being debated. As reported to the FDA, those reactions include:(5)
Headaches/migraines, nausea, abdominal pains, fatigue (blocks sufficient glucose entry into brain), sleep problems, vision problems, anxiety attacks, depression, and asthma/chest tightness.

One common complaint of persons suffering from the effect of aspartame is **memory loss**. Ironically, in 1987, G.D. Searle, the manufacturer of aspartame, undertook a search for a drug to combat memory loss caused by excitatory amino acid damage. Blaylock is one of many scientists and physicians who are concerned about excitatory amino acid damage caused by ingestion of aspartame and MSG. A few of the many experts who have spoken out against the damage being caused by aspartate and glutamate include Adrienne Samuels, Ph.D., an experimental psychologist specializing in research design. Another is Olney, a professor in the department of psychiatry, School of Medicine, Washington University, a neuroscientist and researcher, and one of the world's foremost authorities on excitotoxins. (He informed Searle in 1971 that aspartic acid caused holes in the brain of mice.) Also included is Francis J. Waickman, M.D., a recipient of the Rinkel and Forman Awards, and Board certified in Paediatrics, Allergy, and Immunology.

Other concerned scientists include: John R. Hain, M.D., Board Certified Forensic Pathologist, and H.J. Roberts, M.D., FACP, FCCP, Diabetic Specialist, and selected by a national medical publication as "The Best Doctor in the US"

John Samuels is concerned, also. He compiled a list of scientific research sufficient to show the dangers of ingesting excess free glutamic and aspartic acid.

And there are many more who can be added to this long list.

PHENYLALANINE (50% OF ASPARTAME)

Phenylalanine is an amino acid normally found in the brain. Persons with the genetic disorder, phenylketonuria (PKU) cannot metabolise phenylalanine. This leads to dangerously high levels of phenylalanine in the brain (sometimes lethal). It has been shown that ingesting aspartame, especially along with carbohydrates can lead to excess levels of phenylalanine in the brain even in persons who do not have PKU. **This is not just a theory**, as many people who have eaten large amounts of aspartame over a long period of time and do not have PKU have been shown to have excessive levels of phenylalanine in the blood. Excessive levels of phenylalanine in the brain can cause the levels of serotonin in the brain to decrease, leading to emotional disorders such as **depression**. It was shown in human testing that phenylalanine levels of the blood were increased significantly in human subjects who chronically used aspartame.(6) Even a single use of aspartame raised the blood phenylalanine levels. In his testimony before the US Congress, Dr Louis J. Elsas showed that high blood phenylalanine can be concentrated in parts of the brain, and is especially dangerous for infants and foetuses. He also showed that phenylalanine is metabolised much more efficiently by rodents than by humans.(7)

One account of a case of extremely high phenylalanine levels caused by aspartame was recently published the the "Wednesday Journal" in an article entitled "An Aspartame Nightmare." John Cook began drinking 6 to 8 diet drinks every day. His symptoms started out as memory loss and frequent headaches. He began to crave more aspartame-sweetened drinks. His condition deteriorated so much that he experienced wide mood swings and violent rages. Even though he did not suffer from PKU, a blood test revealed a phenylalanine level of 80 mg/dl. He also showed abnormal brain function and brain damage. After he kicked his aspartame habit, his symptoms improved dramatically. (8)

As Blaylock points out in his book, early studies measuring phenylalanine build-up in the brain were flawed. Investigators who measured specific brain regions and not the average throughout the brain notice significant rises in phenylalanine levels. Specifically the hypothalamus, medulla oblongata, and corpus striatum areas of the brain had the largest increases in phenylalanine. Blaylock goes on to point out that excessive build-up of phenylalanine in the brain can cause schizophrenia or make one more susceptible to seizures.

Therefore, **long-term, excessive use of aspartame may provide a boost to sales of serotonin reuptake inhibitors such as Prozac** and drugs to control schizophrenia and seizures.

METHANOL (AKA WOOD ALCOHOL/POISON) (10% OF ASPARTAME)

Methanol/wood alcohol is a deadly poison. Some people may remember methanol as the poison that has caused some "skid row" alcoholics to end up blind or dead. Methanol is gradually released in the small intestine when the methyl group of aspartame encounter the enzyme chymotrypsin.

The absorption of methanol into the body is sped up considerably when free methanol is ingested. **Free methanol is created from aspartame when it is heated to above 86 Fahrenheit (30 Centigrade). This would occur when aspartame-containing product is improperly stored or when it is heated** (e.g., as part of a "food" product such as Jello).

Methanol breaks down into formic acid and formaldehyde in the body. Formaldehyde is a deadly neurotoxin. An EPA assessment of methanol states that methanol "is considered a cumulative poison due to the low rate of excretion once it is absorbed. In the body, methanol is oxidized to formaldehyde and formic acid; both of these metabolites are toxic." **The recommend a limit of consumption of 7.8 mg/day. A one-litre (approx. 1 quart) aspartame-sweetened beverage contains about 56 mg of methanol. Heavy users of aspartame-containing products consume as much as 250 mg of methanol daily or 32 times the EPA limit. (9)**

Symptoms from methanol poisoning include headaches, ear buzzing, dizziness, nausea, gastrointestinal disturbances, weakness, vertigo, chills, memory lapses, numbness and shooting pains in the extremities, behavioural disturbances, and neuritis. The most well known problems from methanol poisoning are vision problems including misty vision, progressive contraction of visual fields, blurring of vision, obscuration of vision, retinal damage, and blindness. Formaldehyde is a known carcinogen, causes retinal damage, interferes with DNA replication, causes birth defects. (10) Due to the lack of a couple of key enzymes, humans are many times more sensitive to the toxic effects of methanol than animals. Therefore, tests of aspartame or methanol on animals do not accurately reflect the danger for humans. As pointed out by Dr Woodrow C. Monte, Director of the Food Science and Nutrition Laboratory at Arizona State University, "There are no human or mammalian studies to evaluate the possible mutagenic, teratogenic, or carcinogenic effects of chronic administration of methyl alcohol."(11)

He was so concerned about the unresolved safety issues that he filed suit with the FDA requesting a hearing to address these issues. He asked the FDA to "slow down on this soft drink issue long enough to answer some of the important questions. It's not fair that you are leaving the full burden of proof on the few of us who are concerned and have such limited resources. You must remember that you are the American public's last defence. Once you allow usage (of aspartame) there is literally nothing I or my colleagues can do to reverse the course. Aspartame will then join saccharin, the sulfiting agents, and God knows how many other questionable compounds enjoined to insult the human constitution with governmental approval."(10) Shortly thereafter, the Commissioner of the FDA, Arthur Hull Hayes, Jr., approved the use of aspartame in carbonated beverages, he then left for a position with G.D. Searle's Public Relations firm. (11)

It has been pointed out that some fruit juices and alcoholic beverages contain small amounts of methanol. It is important to remember, however, that methanol never appears alone. In every case, ethanol is present, usually in much higher amounts. Ethanol is an antidote for methanol toxicity in humans. (9) The troops of Desert Storm were "treated" to large amounts of aspartame-sweetened beverages which had been heated to over 86 degrees F. in the Saudi Arabian sun. Many of them returned home with numerous disorders similar to what has been seen in persons who have been chemically poisoned by formaldehyde. The free methanol in the beverages may have been a contributing factor in these illnesses. Other breakdown products of aspartame such as DKP (discussed below) may also have been a factor.

In a 1993 act that can only be described as "unconscionable," the FDA approved aspartame as an ingredient in numerous food items that would always be heated to above 86 degrees F (30 degrees C).

DIKETOPIPERAZINE (DKP)

DKP is a by-product of aspartame metabolism. DKP has been implicated in the occurrence of brain tumors. Olney noticed that DKP, when nitrosated in the gut, produced a compound which was similar to N-nitrosourea, a powerful brain tumor causing chemical. Some authors have said that DKP is produced after aspartame ingestion. I am not sure if that is correct. It is definitely true that DKP is formed in liquid aspartame-containing products during prolonged storage.

G.D. Searle conducted animal experiments on the safety of DKP. The FDA found numerous experimental errors occurred, including "clerical errors, mixed-up animals, animals not getting drugs they were supposed to get, pathological specimens lost because of improper handling," and many other errors. (12) These sloppy laboratory procedures may explain why both the test and control animals had sixteen times more brain tumors than would be expected in experiments of this length. In an ironic twist, shortly after these experimental errors were discovered, the FDA used guidelines recommended by G.D. Searle to develop the Industry-wide FDA standards for Good Laboratory Practices. (11) DKP has also been implicated as a cause of uterine polyps and changes in blood cholesterol by FDA Toxicologist Dr Jacqueline Verrett in her testimony before the US Senate. (13)

AILMENTS RESULTING FROM ASPARTAME

The components of aspartame can lead to a wide variety of ailments. Some of these problems occur gradually, others are immediate, acute reactions. There is an enormous population of people who are suffering from symptoms contributed to by aspartame, yet they have no idea why herbs or drugs are not helping relieve their problems. There are other users of aspartame who appear not to be suffering immediate reactions to aspartame. But even these individuals are susceptible to the long-term damage caused by excitatory amino acids, phenylalanine, methanol, and DKP. A few of the many disorders that are of particular concern to me include the following.

Birth Defects.
Dr Diana Dow Edwards, a researcher was funded by Monsanto to study possible birth defects caused by the ingestion of aspartame. After preliminary data showed damaging information about aspartame, funding for the study was cut off. A Genetic Paediatrician at Emory University has testified that aspartame is causing birth defects.7360-367.

In the book, While Waiting: A Prenatal Guidebook by George R. Verrilli, M.D. and Anne Marie Mueser, it is stated that aspartame is suspected of causing brain damage in sensitive individuals. A foetus may be at risk for these effects. Some researchers have suggested that high doses of aspartame may be associated with problems ranging from dizziness and subtle brain changes to mental retardation.

Cancer (Brain Cancer).
In 1981, Satya Dubey, an FDA statistician, stated that the brain tumor data on aspartame was so "worrisome" that he could not recommend approval of NutraSweet. (14) In a two-year study conducted by the manufacturer of aspartame, twelve of the 320 rats fed a normal diet and aspartame developed brain tumors while none of the control rats had tumors. Five of the twelve tumors were in rats given a low dose of aspartame. (15) The approval of aspartame was a violation of the Delaney Amendment which was supposed to prevent cancer-causing substances such as methanol (formaldehyde) and DKP from entering our food supply. The late Dr Adrian Gross, an FDA toxicologist, testified before the US Congress that aspartame was capable of producing brain tumors. This made it illegal for the FDA to set an allowable daily intake at any level. He stated in his testimony that Searle's studies were "to a large extent unreliable" and that "at least one of those studies has established beyond any reasonable doubt that aspartame is capable of inducing brain tumors in experimental animals...." He concluded his testimony by

asking, "What is the reason for the apparent refusal by the FDA to invoke for this food additive the so-called Delaney Amendment to the Food, Drug and Cosmetic Act? .... And if the FDA itself elects to violate the law, who is left to protect the health of the public?"(16)

In the mid-1970s it was discovered that the manufacturer of aspartame falsified studies in several ways. One of the techniques used was to cut tumors out of test animals and put them back in the study. Another technique used to falsify the studies was to list animals that had actually died as surviving the study. Thus, the data on brain tumors was likely worse than discussed above. In addition, a former employee of the manufacturer of aspartame, Raymond Schroeder told the FDA on July 13, 1977 that the particles of DKP were so large that the rats could discriminate between the DKP and their normal diet. (12)

It is interesting to note that the incidence of brain tumors in persons over 65 years of age has increase 67% between the years 1973 and 1990. Brain tumors in all age groups has jumped 10%. The greatest increase has come during the years 1985-1987. (17)

In his book, Aspartame (NutraSweet). Is it Safe? Roberts gives evidence that aspartame can cause a particularly dangerous form of cancer - primary lymphoma of the brain.

Diabetes.
The American Diabetes Association (ADA) is actually recommending this chemical poison to persons with diabetes. According to research conducted by H.J. Roberts, a diabetes specialist, a member of the ADA, and an authority on artificial sweeteners, aspartame:

1) Leads to the precipitation of clinical diabetes.
2) Causes poorer diabetic control in diabetics on insulin or oral drugs.
3) Leads to the aggravation of diabetic complications such as retinopathy, cataracts, neuropathy and gastroparesis.
4) Causes convulsions.

In a statement concerning the use of products containing aspartame by persons with diabetes and hypoglycaemia, Roberts says: "Unfortunately, many patients in my practice, and others seen in consultation, developed serious metabolic, neurologic and other complications that could be specifically attributed to using aspartame products. This was evidenced by:

"The loss of diabetic control, the intensification of hypoglycaemia, the occurrence of presumed 'insulin reactions' (including convulsions) that proved to be aspartame reactions, and the precipitation, aggravation or simulation of diabetic complications (especially impaired vision and neuropathy) while using these products.

"Dramatic improvement of such features after avoiding aspartame, and the prompt predictable recurrence of these problems when the patient resumed aspartame products, knowingly or inadvertently."

Roberts goes on to say:
"I regret the failure of other physicians and the American Diabetes Association (ADA) to sound appropriate warnings to patients and consumers based on these repeated findings which have been described in my corporate-neutral studies and publications."

Blaylock stated that excitotoxins such as that found in aspartame can precipitate diabetes in persons who are genetically susceptible to the disease. (5)

Emotional Disorders.
A double blind study of the effects of aspartame on persons with mood disorders was recently conducted by Dr Ralph G. Walton. **Since the study wasn't funded/controlled by the makers of aspartame, The NutraSweet Company refused to sell him the aspartame.** Walton was forced to obtain and certify it from an outside source.

The study showed a large increase in serious symptoms for persons taking aspartame. Since some of the symptoms were so serious, the Institutional Review Board had to stop the study. Three of the participants had said that they had been "poisoned" by aspartame. Walton concludes that "individuals with mood disorders are particularly sensitive to this artificial sweetener; its use in this population should be discouraged."(18) Aware that the experiment could not be repeated because of the danger to the test subjects, Walton was recently quoted as saying, "I know it causes seizures. I'm convinced also that it definitely causes behavioural changes. I'm very angry that this substance is on the market. I personally question the reliability and validity of any studies funded by the NutraSweet Company."(19)

There are numerous reported cases of low brain serotonin levels, depression and other emotional disorders that have been linked to aspartame and often are relieved by stopping the intake of aspartame.

Researchers have pointed out that increasing in phenylalanine levels in the brain, which can and does occur in persons without PKU, leads to a decreased level of the neurotransmitter, serotonin, which leads to a variety of emotional disorders. Dr William M. Pardridge of UCLA testified before the US Senate that a youth drinking four 16-ounce bottles of diet soda per day leads to an enormous increase in the phenylalanine level.

Epilepsy/Seizures.

With the large and growing number of seizures caused by aspartame, it is sad to see that the Epilepsy Foundation is promoting the "safety" of aspartame. At Massachusetts Institute of Technology, 80 people who had suffered seizures after ingesting aspartame were surveyed. Community Nutrition Institute concluded the following about the survey:

"These 80 cases meet the FDA's own definition of an imminent hazard to the public health, which requires the FDA to expeditiously remove a product from the market."

Both the Air Force's magazine Flying Safety and the Navy's magazine, Navy Physiology published articles warning about the many dangers of aspartame including the cumulative deleterious effects of methanol and the greater likelihood of birth defects. The articles note that the ingestion of aspartame can make pilots more susceptible to seizures and vertigo. Twenty articles sounding warnings about ingesting aspartame while flying have also appeared in the National Business Aircraft Association Digest (NBAA Digest 1993), Aviation Medical Bulletin (1988), The Aviation Consumer (1988), Canadian General Aviation News (1990), Pacific Flyer (1988), General Aviation News (1989), Aviation Safety Digest (1989), and Plane and Pilot (1990) and a paper warning about aspartame was presented at the 57th Annual Meeting of the Aerospace Medical Association (Gaffney 1986).

Recently, a hotline was set up for pilots suffering from acute reactions to aspartame ingestion. Over 600 pilots have reported symptoms including some who have reported suffering grand mal seizures in the cockpit due to aspartame. (21)

One of the original studies on aspartame was performed in 1969 by an independent scientist, Dr Harry Waisman. He studied the effects of aspartame on infant primates. Out of the seven infant monkeys, one died after 300 days and five others had grand mal seizures. Of course, these negative findings were not submitted to the FDA during the approval process. (22)

Why don't we hear about these things?

The reason many people do not hear about serious reactions to aspartame is twofold:
1) Lack of awareness by the general population. Aspartame-caused diseases are not reported in the newspapers like plane crashes. This is because these incidents occur one at a time in thousands of different locations across the US.
2) Most people do not associate their symptoms with the long-term use of aspartame. For the people who have killed a significant percentage of the brain cells and thereby caused a chronic illness, there is no way that they would normally associate such an illness with aspartame consumption. **How aspartame was approved is a lesson in how chemical and pharmaceutical companies can manipulate government agencies such as the FDA, "bribe" organizations such as the American Dietetic Association, and flood the scientific community with flawed and fraudulent industry-sponsored studies funded by the makers of aspartame.**

Erik Millstone, a researcher at the Science Policy Research Unit of Sussex University has compiled thousands of pages of evidence, some of which have been obtained using the freedom of information act 23, showing:

1. Laboratory tests were faked and dangers were concealed.
2. Tumors were removed from animals and animals that had died were "restored to life" in laboratory records.
3. False and misleading statements were made to the FDA.
4. The two US Attorneys given the task of bringing fraud charges against the aspartame manufacturer took positions with the manufacturer's law firm, letting the statute of limitations run out.
5. The Commissioner of the FDA overruled the objections of the FDA's own scientific board of inquiry. Shortly after that decision, he took a position with Burson-Marsteller, the firm in charge of public relations for G.D. Searle.

A Public Board of Inquiry (PBOI) was conducted in 1980. There were three scientists who reviewed the objections of Olney and Turner to the approval of aspartame. They voted unanimously against aspartame's approval. The FDA Commissioner, Dr Arthur Hull Hayes, Jr. then created a 5-person Scientific Commission to review the PBOI findings. After it became clear that the Commission would uphold the PBOI's decision by a vote of 3 to 2, another person was added to the Commission, creating a deadlocked vote. This allowed the FDA Commissioner to break the deadlock and approve aspartame for dry goods in 1981. Dr Jacqueline Verrett, the Senior Scientist in an FDA Bureau of Foods review team created in August 1977 to review the Bressler Report (a report that detailed G.D. Searle's abuses during the pre-approval testing) said:

"It was pretty obvious that somewhere along the line, the bureau officials were working up to a whitewash." In 1987, Verrett testified before the US Senate stating that the experiments conducted by Searle were a "disaster." She stated that her team was instructed not to comment on or be concerned with the overall validity of the studies. She stated that questions about birth defects have not been answered. She continued her testimony by discussing the fact that DKP has been shown to increase uterine polyps and change blood cholesterol and that increasing the temperature of the product leads to an increase in production of DKP. (13)

Revolving doors

The FDA and the manufacturers of aspartame have had a revolving door of employment for many years. In addition to the FDA Commissioner and two US Attorneys leaving to take positions with companies connected with G.D. Searle, four other FDA officials connected with the approval of aspartame took positions connected with the NutraSweet industry between 1979 and 1982 including the Deputy FDA Commissioner, the Special Assistant to the FDA Commissioner, the Associate Director of the Bureau of Foods and Toxicology and the Attorney involved with the Public Board of Inquiry. (24)

It is important to realize that this type of revolving-door activity has been going on for decades. The Townsend Letter for Doctors (11/92) reported on a study revealing that 37 of 49 top FDA officials who left the FDA took positions with companies they had regulated. They also reported that over 150 FDA officials owned stock in drug companies they were assigned to manage. Many organizations and universities receive large sums of money from companies connected to the NutraSweet Association, a group of companies promoting the use of aspartame. In January 1993, the American Dietetic Association received a US$75,000 grant from the NutraSweet Company. The American Dietetic Association has stated that the NutraSweet Company writes their "Facts" sheets. (25)

Many other "independent" organizations and researchers receive large sums of money from the manufacturers of aspartame. The American Diabetes Association has received a large amount of money from Nutrasweet, including money to run a cooking school in Chicago (presumably to teach diabetes how to use Nutrasweet in their cooking).

A researcher in New England who has pointed out the dangers of aspartame in the past is now a Monsanto consultant. Another researcher in the South-eastern US had testified about the dangers of aspartame on foetuses. An investigative reporter has discovered that he was told to keep his mouth shut to avoid causing the loss of a large grant from a diet cola manufacturer in the NutraSweet Association.

What is the FDA doing to protect the consumer from the dangers of aspartame? Less than nothing.

In 1992, the FDA approved aspartame for use in malt beverages, breakfast cereals, and refrigerated puddings and fillings. In 1993 the FDA approved aspartame for use in hard and soft candies, non-alcoholic favored beverages, tea beverages, fruit juices and concentrates, baked goods and baking mixes, and frostings, toppings and fillings for baked goods.

In 1991, the FDA banned the importation of stevia. The powder of the leaf has been used for hundreds of years as an alternative sweetener. It is used widely in Japan with no adverse effects. Scientists involved in reviewing stevia have declared it to be safe for human consumption - something which has been well known in many parts of the world where it is not banned. Everyone that I have spoken with in regards to this issue believes that stevia was banned to keep the product from taking hold in the US and cutting into sales of aspartame.(26)

What is the US Congress doing to protect the consumer from the dangers of aspartame? Nothing.

What is the US Administration (President) doing to protect the consumer from the dangers of aspartame? Nothing.

Aspartame consumption is not only a problem in the US. It is being sold in over 70 countries throughout the world.

ASPARTAME CAN BE FOUND IN:
- Instant breakfasts
- Breath mints
- Cereals
- Sugar-free chewing gum
- Cocoa mixes
- Coffee beverages
- Frozen desserts
- Gelatine desserts
- Juice beverages
- Laxatives
- Multivitamins
- Milk drinks
- Pharmaceuticals and supplements
- Shake mixes
- Soft drinks
- Tabletop sweeteners
- Tea beverages
- Instant teas and coffees
- Topping mixes
- Wine coolers
- Yoghurt

I have been told that aspartame has been found in products where it is not listed on the label. One must be particular careful of pharmaceuticals and supplements. I have been informed that even some supplements made by well-known supplement manufacturers such as Twinlabs contain aspartame.

The information I have related above is just the tip of the iceberg as far as damaging information about aspartame. In order for the reader to find out more, I have included some resources below.

BOOKS:

   * Blaylock, Russell L., Excitotoxins: The Taste That Kills (Health Press, Santa Fe, New Mexico, c1994). One of the best books available on excitotoxins. Well worth reading!
   * H. J. Roberts, M.D., Aspartame (NutraSweet), Is it Safe? Available from the Aspartame Consumer Safety Network.
   * Sweet'ner Dearest, Available from the Aspartame Consumer Safety Network

* Mary Nash Stoddard, The Deadly Deception, Available from the Aspartame Consumer Safety Network.
* Barbara Mullarkey, Editor, Bittersweet Aspartame - A Diet Delusion,
* Available from the Aspartame Consumer Safety Network.
* The Aspartame Consumer Safety Network, The Aspartame Consumer Safety Network Synopsis.
* Dennis Remington, M.D. and Barbara Higa, R.D., The Bitter Truth About Artificial Sweetners, Available from the Aspartame Consumer Safety Network

ASPARTAME CONSUMER SAFETY NETWORK
PO Box 780634
Dallas, Texas 75378, USA.
Phone: (214) 352-4268

REFERENCES

(1) Department of Health and Human Services, Report on All Adverse Reactions in the Adverse Reaction Monitoring System, (February 25 and 28, 1994).
(2) Compiled by researchers, physicians, and artificial sweetener experts for Mission Possible, a group dedicated to warning consumers about aspartame.
(3) Excitotoxins: The Taste That Kills, by Russell L. Blaylock, M.D.
(4) Safety of Amino Acids, Life Sciences Research Office, FASEB, FDA Contract No. 223-88-2124, Task Order No. 8.
(5) FDA Adverse Reaction Monitoring System.
(6) Wurtman and Walker, "Dietary Phenylalanine and Brain Function," Proceedings of the First International Meeting on Dietary Phenylalanine and Brain Function., Washington, D.C., May 8, 1987.
(7) Hearing Before the Committee On Labor and Human Resources United States Senate, First Session on Examining the Health and Safety Concerns of Nutrasweet (Aspartame).
(8) Account of John Cook as published in Informed Consent Magazine. "How Safe Is Your Artificial Sweetner" by Barbara Mullarkey, September/October 1994.
(9) Woodrow C. Monte, Ph.D., R.D., "Aspartame: Methanol and the Public Health," Journal of Applied Nutrition, 36 (1): 42-53.
(10) US Court of Appeals for the District of Columbia Circuit, No. 84-1153 Community Nutrition Institute and Dr Woodrow Monte v. Dr Mark Novitch, Acting Commissioner, US FDA (9/24/85).

(11) Aspartame Time Line by Barbara Mullarkey as published in Informed Consent Magazine, May/June 1994.

(12) FDA Searle Investigation Task Force. "Final Report of Investigation of G.D. Searle Company." (March 24, 1976)

(13) Testimony of Dr Jacqueline Verrett, FDA Toxicologist before the US Senate Committee on Labor and Human Resources, (November 3, 1987).

(14) Internal FDA memorandum.

(15) Analysis prepared by Dr John Olney as a statement before the Aspartame Board of Inquire of the FDA. Also Excitotoxins by Russell Blaylock, M.D.

(16) Congressional Record SID835: 131 (August 1, 1985)

(17) National Cancer Institute SEER Program Data.

(18) Walton, Ralph G., Robert Hudak, Ruth Green-Waite "Adverse Reactions to Aspartame: Double-Blind Challenge in Patients from a Vulnerable Population," Biological Psychiatry, 1993:34:13-17.

(19) Barbara Mullarkey, "How Safe Is Your Artificial Sweetner," September/October 1994 issue of Informed Consent Magazine.

(20) US Air Force. "Aspartame Alert." Flying Safety, 48 (5): 20-21 (May 1992).

(21) Reported by the Aspartame Consumer Safety Network.

(22) Barbara Mullarkey, Bittersweet Aspartame, A Diet Delusion.

(23) Millstone, Eric "Sweet and Sour." The Ecologist, 25 (March/April 1994).

(24) Mary Nash Stoddard, Editor, "The Deadly Deception," Aspartame Consumer Safety Network.

(25) ADA Courier, January 1993, Volume 32, Number 1. (26) "FDA Rejects AHPA Stevia Petition" by Mark Blumenthal, Whole Foods, April 1994.

Hi All,

I'm sure you're already aware of some of the side affects of these poisons, thought I'd forward this article as it makes some disturbing observations.

Regards,
Cathy

Subject: ...Aspartame....

THE MULTIPLE SCLEROSIS FOUNDATION & FDA ARE SUING FOR COLLUSION WITH MONSANTO - Article written by Betty Martini

I have spent several days lecturing at the World Environmental Conference on "Aspartame: Marketed as 'NutraSweet', 'Equal', and 'Spoonful'". In the keynote address by the EPA, they announced that there was an epidemic of multiple sclerosis and systemic lupus, and they did not understand what toxin was causing this to be rampant across the United States. I explained I was there to lecture on exactly that subject.

When the temperature of aspartame exceeds 86 degrees F, the wood alcohol in aspartame coverts to formaldehyde and then to formic acid, which in turn causes metabolic acidosis. (Formic acid is the poison found in the sting of fire ants.) The methanol toxicity mimics multiple sclerosis; thus, people were being diagnosed with having multiple sclerosis in error. The multiple sclerosis is not a death sentence, where methanol toxicity is.

In the case of systemic lupus, we are finding it has become almost as rampant as multiple sclerosis, especially in Diet Coke and Diet Pepsi drinkers. Also, with methanol toxicity, the victims usually drink three to four 12-oz. cans of them per day, some even more. In the cases of systemic lupus, which is triggered by aspartame, the victim usually does not know that the aspartame is the culprit. The victim continues its use aggravating the lupus to such a degree, that sometimes it becomes life threatening.

When we get people off the aspartame, those with systemic lupus usually become asymptomatic. Unfortunately, we cannot reverse this disease. On the other hand, in the case of those diagnosed with Multiple Sclerosis (when in reality, the disease is methanol toxicity), most of the symptoms disappear. We have seen cases where their vision has returned and even their hearing has returned. This also applies to cases of tinnitus.

During a lecture I said, "If you are using aspartame [NutraSweet, Equal, Spoonful, etc.] and you suffer from fibromyalgia symptoms, spasms, shooting pains, numbness in your legs, cramps, vertigo, dizziness, headaches, tinnitus, joint

pain, depression, anxiety attacks, slurred speech, blurred
vision, or memory loss, you probably have ASPARTAME DISEASE!"
People were jumping up during the lecture saying, "I've got
this! Is it reversible?" It is rampant. Some of the speakers
at my lecture even were suffering from these symptoms.

In one lecture attended by the Ambassador of Uganda, he told us
that their sugar industry is adding aspartame! He continued by
saying that one of the industry leader's son could no longer
walk due in part to product usage! We have a very serious
problem.

Even a stranger came up to Dr. Espisto, (one of my speakers)
and myself and said, "Could you tell me why so many people seem
to be coming down with MS?" During a visit to a hospice, a
nurse said that six of her friends, who were heavy Diet Coke
drinkers, had all been diagnosed with MS. This is beyond
coincidence.

Here is the problem. There were Congressional Hearings when
aspartame was originally included as a sweetener in 100
different products. Since this initial hearing, there have
been two subsequent hearings, but to no avail. Nothing has been
done. The drug and chemical lobbies have very deep pockets. Now
there are over 5,000 products containing this chemical, and the
PATENT HAS EXPIRED!!

At the time of this first hearing, people were going blind.
The methanol in the Aspartame converts to formaldehyde in the
retina of the eye. Formaldehyde is grouped in the same class of
drugs as cyanide and arsenic - DEADLY POISONS!!! Unfortunately,
it just takes longer to quietly kill, but it is killing people
and causing all kinds of neurological problems.

Aspartame changes the brain's chemistry. It is the reason for
severe seizures. This drug changes the dopamine level in the
brain. Imagine what this drug does to patients suffering from
Parkinson's Disease. This drug also causes birth defects.
There is absolutely no reason to take this product.

It is NOT A DIET PRODUCT!! The Congressional record said, "It
makes you crave Carbohydrates and will make you FAT." Dr.
Roberts stated that when he got patients off aspartame, their

average weight loss was 19 pounds per person. The formaldehyde stores in the fat cells, particularly in the hips and thighs.

Aspartame is especially deadly for diabetics. All physicians know what wood alcohol will do to a diabetic. We find that physicians believe that they have patients with retinopathy, when in fact, it is caused by the aspartame. The aspartame keeps the blood sugar level out of control, causing many patients to go into a coma. Unfortunately, many have died.

People were telling us at the conference of the American College of Physicians that they had relatives that had switched from saccharin to an aspartame product and how that relative had eventually gone into a coma. Their physicians could not get the blood sugar levels under control; thus the patients suffered acute memory loss and eventually coma and death.

Memory loss is due the fact that aspartic acid and phenylalanine are neurotoxic without the other amino acids found in protein, thus it goes past the blood brain barrier and deteriorates the neurons of the brain. Dr. Russell Blaylock, a prominent neurosurgeon of Jackson, Mississippi, said, "The ingredients stimulate the neurons of the brain to death, causing brain damage of varying degrees." Dr. Blaylock has written a book entitled EXCITOTOXINS: The Taste That Kills. (Health Press 1-800-643-2665).

Dr. H. J. Roberts, diabetic specialist and world expert on aspartame poisoning, has also written a book entitled Defense Against Alzheimer's Disease (1-800-814-9800). Dr. Roberts tells how aspartame poisoning is escalating Alzheimer's Disease, and indeed it is. As the hospice nurse told me, women are being admitted at 30 years of age with Alzheimer's Disease.

Dr. Blaylock and Dr. Roberts will be writing a position paper with some case histories and will post it on the Internet. According to the Conference of the American College of Physicians, "We are talking about a plague of neurological diseases caused by this deadly poison."

Dr. Roberts realized what was happening when aspartame was first marketed. He said, "His diabetic patients presented memory loss, confusion, and severe vision loss." At the Conference of the American College of Physicians, doctors admitted that they did not know. They had wondered why seizures were rampant (the phenylalanine in aspartame breaks down the seizure threshold and depletes serotonin, which causes manic depression, panic attacks, rage and violence).

Just before the Conference, I received a fax from Norway asking for a possible antidote for this poison because they are experiencing so many problems in their country. This "poison" is now available in 90 PLUS countries worldwide. Fortunately, we had speakers and ambassadors at the conference from different nations who have pledged their help. We ask that you help too.

Print this article out and warn everyone you know. Take anything that contains aspartame back to the store. Take the "NO ASPARTAME TEST" and send us your case history.

I assure you that MONSANTO, the creator of aspartame, knows how deadly it is. They fund the American Medical Association, American Dietetic Association, Congress, and the Conference of the American College of Physicians. The New York Times, November 15, 1996, ran an article on how the American Dietetic Association takes money from the food industry to endorse their products. Therefore, they cannot criticize any additives or tell about their link to MONSANTO.

How bad is this? We told a mother who had a child on NutraSweet to get off the product. The child was having grand mal seizures every day. The mother called her physician, who called the ADA, who told the doctor not to take the child off the NutraSweet. We are still trying to convince the mother that the aspartame is causing the seizures. Every time we get someone off of aspartame, the seizures stop. If the baby dies, you know whose fault it is, and what we are up against.

There are 92 documented symptoms of aspartame, from coma to death. The majority of them are all neurological, because the aspartame destroys the nervous system. Aspartame Disease is partially the cause to what is behind some of the mystery of the Desert Storm health problems. The burning tongue and other problems discussed in over 60 cases can be directly related to the consumption of aspartame product. Several thousand pallets

of diet drinks were shipped to the Desert Storm troops.
(Remember heat can liberate the methanol from the aspartame at
86 degrees F.) Diet drinks sat in the 120 degree F Arabian sun
for weeks at a time on pallets. The servicemen and women drank
them all day long. All of their symptoms are identical to
aspartame poisoning.

Dr. Roberts says "Consuming aspartame at the time of conception
can cause birth defects." According to Dr. Louis Elsas,
Pediatrician and Professor of Genetics, at Emory University, in
his testimony before Congress, " The phenylalanine concentrates
in the placenta causing mental retardation. In the original
lab tests, animals developed brain tumors, phenylalanine breaks
down into DXP (a brain tumor agent.) When Dr. Espisto was
lecturing on aspartame, one physician in the audience, a
neurosurgeon, said, "When they remove brain tumors, they have
found high levels of aspartame in them."

Although Stevia, a sweet food, NOT AN ADDITIVE, which helps in
the metabolism of sugar, which would be ideal for diabetics,
has now been approved as a dietary supplement by the FDA for
years, the FDA has outlawed this sweet food because of their
loyalty to MONSANTO. If it says "SUGAR FREE" on the label - DO
NOT EVEN THINK ABOUT IT.

Senator Howard Hetzenbaum wrote a bill that would have warned
all infants, pregnant mothers and children of the dangers of
aspartame. The bill would have also instituted independent
studies on the problems existing in the population (seizures,
changes in brain chemistry, changes in neurological and
behavioral symptoms). It was killed by the powerful drug and
chemical lobbies, letting loose the hounds of disease and death
on an unsuspecting public. Since the Conference of the
American College of Physicians, we hope to have the help of
several world leaders.

Again, please help us, too. There are a lot of people out
there who must be warned, please let them know this information.

Women's Cancer Resource Center Laurie Moser, Assistant Director
1815 East 41st Street, Suite C Minneapolis, MN 55407-3425
1-800-908-8544 or 612-729-049

513-VACCINES
"Your failure to be informed does not make me a wacko." -- John Loeffler
I have no medical training, but I can read.

I find this topic severely distressing. Even those words fail to convey my feelings. Aggravating that distress is that I am not sure if it is simply a case of human madness and insanity initiated and motivated by callous greed, which would be the "best" case scenario, or if in fact there is something far more sinister than mere corporate greed implementing and sustaining this assault on humanity.

If vaccinations were an isolated or "stand alone" oddity, I may find it as just that, but as there are simultaneously literally hundreds of other similarly debilitating assaults on humanity, I am forced to doubt that it is more than an unfortunate set of circumstances. Pardon me if I see a pattern where others see nothing unusual, but hundreds of similar "hits" tend to make one look for a common cause. (This may be an ability identified by an intelligence test that identified me as a visual mathematician with ability to see patterns and make correlations,)

Unfortunately when it comes down to the debilitation or death of humanity, a common source is actually identifiable. Our commonly accepted "god" is obviously not only not helping us, but complicit. Writings in other sections of this book will give validation and also all the justification of this opinion. Jehovah or Enlil and the heirs are not sympathetic to humanity. God I hate it when a god is not on "our" side. Do the research. (The history of the "Genesis God" is elsewhere in this book.) It should become obvious that some non-human 3rd party is the author of unending schemes that should be obviously designed to create misery or destroy.

There can be absolutely no doubt, when the facts and case histories are laid open and exposed, that untold deaths, debilitation, and injury are a direct result of the vaccination "industry".

Regardless of what I think, it is imperative that you, the reader, put yourself in a position where you really can make an informed decision about the entire subject rather than just blindly accept and submit yourself and family to this practice.

Death and permanent injury are not denied by the "industry", rather they play down the unfortunates numbers and statistics as **just bad luck**, or a result of **"complications"** that are **unfortunate**. Put it this way: would you join a queue knowing at every so often one of you would be randomly taken out for immediate execution? Would you put your children in that line? Would you allow yourself or family to play with a loaded revolver?

As this is indeed such a painful issue (and I have had family issues relating to "bad reaction" to vaccines that resulted in our sane family doctor advising "no more vaccinations for the child") I must personally be brief in my comments. However the matter requires your most urgent attention and investigations. Hereunder are some pointers and information that are merely a start. The resources available for your consideration are larger than you can imagine.

In short we the people, are being wilfully and knowingly killed off. No one is being held accountable for what is a genocide. Let's put aside emotional writing and get down to history.

****

1963. The mass **vaccination** campaigns of the 1950s and '60s may be causing hundreds of deaths a year because of a cancer-causing virus that contaminated the first polio vaccine, according to scientists. Known as SV40, the virus came from dead monkeys whose kidney cells were used to culture the first Salk vaccines. (Nice hey? – Bob) Doctors estimate that the virus was injected into tens of millions during the vaccination campaigns, including several million in Canada, **before being detected and screened out in 1963.** Those born between 1941 and 1961 are thought to be most at risk of having been infected.

Mid-1970's. The incidence of AIDS infections in Africa coincides exactly with the locations of the W.H.O. smallpox vaccination program in the mid-1970's (London Times, May 11, 1987). Some 14,000 Haitians then on UN secondment to Central Africa were also vaccinated in this campaign. Personnel actually conducting the vaccinations may have been completely unaware that the vaccine was anything other than what they were told.

1978. The Hepatitis B vaccine study appears to have been the initial means of planting the infection in New York City. The test protocol specified **non-monogamous males** only, and **homosexuals received a different vaccine from heterosexuals**. At least 25-50% of the first reported New York AIDS cases in 1981 had received the Hepatitis B test vaccine in 1978. **By 1984, 64% of the vaccine recipients had AIDS**, and the figures on the current infection rate for the participants of that study are held by the U.S. Department of Justice, and "unavailable."

****

"As a legislator, I believe mandated smallpox vaccines are very bad policy. The point is not that smallpox vaccines are necessarily a bad idea, but rather that intimately personal medical decisions should not be made by government. The real issue is individual medical choice. No single person, including the President of the United States, should ever be given the power to make a medical decision for potentially millions of Americans. Freedom over one's physical person is the most basic freedom of all, and people in a free society should be sovereign over their own bodies. When we give government the power to make medical decisions for us, we in essence accept that the state owns our bodies." (Ron Paul MD, LewRockwell.com)

Vaccinations - good or bad? updated April 2008

The idea of vaccination is that if you give the immune system a small "taste" of a bug (such as polio, whooping cough etc) it will make antibodies which will protect one against future exposures to the real thing. Good idea, but in practice it is not as simple.

My medical training tells me that all these issues should be resolved by logical argument. But in the modern world, all these arguments are tainted by **vested interest (primarily from drug companies)** and it is difficult to trust the data with which one is presented. Therefore one ends up working from either limited data, or untrustworthy data, or common sense and experience and ends up with a belief. So what you are getting below are my individual conclusions, but it is up to every parent to find what information they can and make up their own minds.

The evidence that vaccinations reduce incidence of disease is pretty thin. **Most infectious diseases have declined as a result of improved hygiene and nutrition.** Doctors believe that vaccinations work and are reluctant to diagnose a disease in a vaccinated child. **So for example since polio vaccine, polio is rarely diagnosed, but there has been an increase in aseptic meningitis.**

The medical profession, backed up by the pharmaceutical and chemical industry, are experts in cover-ups. When doctors find themselves in trouble they close ranks. Most people have seen cover-ups for themselves with **drug side effects (which kill huge numbers of people every year but are hushed up).** I see cover ups in patients with pesticide poisoning, with problems from silicone breast implants and in Gulf War syndrome. Doubtless there are others and I know **vaccine damage is covered up and/or denied. I have seen too many children with serious health problems dating from vaccination for which there is no other explanation for their illness.** I have to believe the evidence of my own eyes.

Vaccines can cause harm

There is now strong evidence that part of Gulf War Syndrome was caused by multiple vaccinations. MMR has been linked with autism and there is still a case to be answered here. There are many cases of brain damaged children following triple vaccine (diptheria, pertussis, tetanus).

Vaccines may be causing harm in unseen ways

Polio vaccination may be the cause of the huge increase in post viral fatigue syndrome. Before polio vaccination, post viral syndrome was rare. This is because people caught polio (which occasionally results in paralysis) which is an enterovirus. They mounted a vigorous immune reaction against polio virus which gave them cross-immunity against all other enteroviruses including Epstein Barr (glandular fever), coxsackie B, ECHO etc. This protected them against post viral fatigue since this most commonly follows an enteroviral infection.

We now know that many cancers are caused by viral infection. Obvious examples include hepatitis B (primary liver cancer), cervical wart virus (cervical cancer) and AIDS (Kaposi's sarcoma). Chronic myeloid leukaemia is probably virally induced. How many other cancers could there be from which we are protected by proper exposure to a virus, but not protected by vaccination?

Nobody knows the answer to this question. And certainly no studies are being done.

What is in a vaccine?

Not just bits of bacteria and viruses. No immune system is going to react vigorously against a few dead or half alive (attenuated) cells. To turn the immune system on a vaccine needs an immune adjuvant added. These include aluminium and mercury which are toxic in their own right. It may well be that autism following MMR is actually a mercury problem.

Vaccines are made from bugs which are grown in animal tissues including beef. There is evidence to suggest that the cases of new variant CJD in young people may be due to direct injection of prion from these tissue cultures.

So what is the alternative?

We should be tackling infectious disease by good hygiene and boosting the immune system.

Good hygiene

By good hygiene definitely I do not mean obsessively wiping down working surfaces with antiseptic wipes. Indeed this is counter productive because we need daily exposure to bacteria to train and programme the immune system. What I recommend is proper public health measures such as:

* Not pumping raw sewage into the seas for people to swim in.
* Not making animals travel hundreds of miles to slaughter houses so they crap themselves on the way and get covered with shit, contaminating meat subsequently. Please try to buy local produce, or organic produce which has animal care standards.
* Not keeping chickens so intensively that they need constant antibiotics to survive chronic salmonella.
* Moving towards more organic farming practices i.e. away from intensive farming, use of properly composted animal waste, using local suppliers etc.
* Sexually transmitted diseases are presently all too common. Take proper precautions.
* We should not be concentrating sick and ill people in large general hospitals. This means that antibiotic resistant organisms can develop and spread quickly from one patient to the next.

* There are many other ideas for good hygiene. It is important to think carefully for yourselves. Please do not assume that "hygienic" chemical solutions are the answer.

Boosting the immune system

Human beings live on a knife edge with their immune systems. The immune system has a delicate balancing act because it needs to be able to recognise bugs and attack them, it must recognise cancer cells and attack them, but it must recognise "self" i.e. human bacteria/bugs and human cells and and ignore them. It is already confused by chemicals.

We should not be thinking about getting rid of the bug. This will always be impossible simply because "nature abhors a vacuum" and if you get rid of one bug, another will take its place.

Therefore we should be thinking about individual resistance to disease i.e. making people so healthy through good diet, good micronutrient status (vitamins and minerals) and freedom from toxins (i.e. red herrings and obstacles) that the immune system can easily resist any bugs that do gain entry. For example, measles can cause eye damage, but not if there is good vitamin A status.

The trouble is that against all these arguments is the combined weight of the medical profession and pharmaceutical companies who financially drive government and control the Press telling us that vaccination is safe and desirable. Nowadays logical argument no longer prevails and policy is dictated by big business and cash.

So what vaccinations would I give my child today?

No DPT in the first few months of life (I would look for protection from breast feeding).

I would give polio because I do not want to risk paralysis and I believe good nutrition will protect my child from other severe enteroviral infections.

No MMR (I want my child to get these infections young when the immune system, with good nutrition, can deal efficiently with these infections). See MMR vaccination - should my child have it?

Good HealthKeeping is a website with information on obtaining single vaccines.
Once my child started running around outside I would give tetanus vaccination.

No BCG.

With a daughter, I would check rubella status as a teenager and use vaccination if she was negative: I do not want her to get rubella during pregnancy.

In conclusion
These, as I say, are my beliefs. They may well change in the future as I learn new things.

For a list of vaccine ingredients refer to RENSE.COM

Rense.com

Vaccine Ingredients -
Formaldehyde, Aspartame,
Mercury, Etc
11-11-4

This following list of common vaccines and their ingredients should shock anyone.

The numbers of microbes, antibiotics, chemicals, heavy metals and animal byproducts is staggering. Would you knowingly inject these materials into your children?

Acel-Immune DTaP - **Diphtheria-Tetanus-Pertussis** Wyeth-Ayerst 800.934.5556
* diphtheria and tetanus toxoids and acellular pertussis adsorbed, formaldehyde, aluminum hydroxide, aluminum phosphate, thimerosal, and polysorbate 80 (Tween-80) gelatin Act HIB

Haemophilus - **Influenza B** Connaught Laboratories 800.822.2463
* Haemophilus influenza Type B, polyribosylribitol phosphate ammonium sulfate, formalin, and sucrose

Attenuvax - **Measles** Merck & Co., Inc. 800-672-6372

* measles live virus neomycin sorbitol hydrolized gelatin, chick embryo

Biavax - **Rubella** Merck & Co., Inc. 800-672-6372
* rubella live virus neomycin sorbitol hydrolized gelatin, human diploid cells from aborted fetal tissue

BioThrax - Anthrax Adsorbed BioPort Corporation 517.327.1500
* nonencapsulated strain of Bacillus anthracis aluminum hydroxide, benzethonium chloride, and formaldehyde

DPT - **Diphtheria-Tetanus-Pertussis** GlaxoSmithKline 800.366.8900 x5231
* diphtheria and tetanus toxoids and acellular pertussis adsorbed, formaldehyde, aluminum phosphate, ammonium sulfate, and thimerosal, washed sheep RBCs

Dryvax - **Smallpox** (not licensed d/t expiration) Wyeth-Ayerst 800.934.5556
* live vaccinia virus, with "some microbial contaminants," according to the Working Group on Civilian Biodefense polymyxcin B sulfate, streptomycin sulfate, chlortetracycline hydrochloride, and neomycin sulfate glycerin, and phenol -a compound obtained by distillation of coal tar vesicle fluid from calf skins Engerix-B

Recombinant **Hepatitis B** GlaxoSmithKline 800.366.8900 x5231
* genetic sequence of the hepatitis B virus that codes for the surface antigen (HbSAg), cloned into GMO yeast, aluminum hydroxide, and thimerosal

Fluvirin Medeva Pharmaceuticals 888.MEDEVA 716.274.5300
* influenza virus, neomycin, polymyxin, beta-propiolactone, chick embryonic fluid

FluShield Wyeth-Ayerst 800.934.5556
* trivalent influenza virus, types A&B gentamicin sulphate formadehyde, thimerosal, and polysorbate 80 (Tween-80) chick embryonic fluid

Havrix - **Hepatitis A** GlaxoSmithKline 800.366.8900 x5231
* hepatitis A virus, formalin, aluminum hydroxide, 2-phenoxyethanol, and polysorbate 20 residual MRC5 proteins - human diploid cells from aborted fetal tissue

HiB Titer - Haemophilus **Influenza B** Wyeth-Ayerst 800.934.5556
* haemophilus influenza B, polyribosylribitol phosphate, yeast, ammonium sulfate, thimerosal, and chemically defined yeast-based medium

Imovax Connaught Laboratories 800.822.2463

* rabies virus adsorbed, neomycin sulfate, phenol, red indicator human albumin, human diploid cells from aborted fetal tissue

IPOL Connaught Laboratories 800.822.2463
* 3 types of **polio** viruses neomycin, streptomycin, and polymyxin B formaldehyde, and 2-phenoxyethenol continuous line of monkey kidney cells

JE-VAX - Japanese Ancephalitis Aventis Pasteur USA 800.VACCINE
* Nakayama-NIH strain of Japanese encephalitis virus, inactivated formaldehyde, polysorbate 80 (Tween-80), and thimerosal mouse serum proteins, and gelatin

LYMErix - Lyme GlaxoSmithKline 888-825-5249
* recombinant protein (OspA) from the outer surface of the spirochete Borrelia burgdorferi kanamycin aluminum hydroxide, 2-phenoxyethenol, phosphate buffered saline

MMR - **Measles-Mumps-Rubella** Merck & Co., Inc. 800.672.6372
* measles, mumps, rubella live virus, neomycin sorbitol, hydrolized gelatin, chick embryonic fluid, and human diploid cells from aborted fetal tissue

M-R-Vax - **Measles-Rubella** Merck & Co., Inc. 800.672.6372
* measles, rubella live virus neomycin sorbitol hydrolized gelatin, chick embryonic fluid, and human diploid cells from aborted fetal tissue

Menomune - **Meningococcal** Connaught Laboratories 800.822.2463
* freeze-dried polysaccharide antigens from Neisseria meningitidis bacteria, thimerosal, and lactose

Meruvax I - **Mumps** Merck & Co., Inc. 800.672.6372
* mumps live virus neomycin sorbitol hydrolized gelatin

NYVAC - (new **smallpox** batch, not licensed) Aventis Pasteur USA 800.VACCINE
* highly-attenuated vaccinia virus, polymyxcin   B, sulfate, streptomycin   sulfate,   chlortetracycline   hydrochloride,   and neomycin sulfate glycerin, and phenol -a compound obtained by distillation of   coal tar vesicle fluid from calf skins

Orimune - **Oral Polio** Wyeth-Ayerst 800.934.5556
* 3 types of polio viruses, attenuated neomycin, streptomycin sorbitol monkey kidney cells and calf serum

Pneumovax - **Streptococcus** Pneumoniae Merck & Co., Inc. 800.672.6372
* capsular polysaccharides from polyvalent (23 types), pneumococcal bacteria, phenol,

Prevnar **Pneumococcal** - 7-Valent Conjugate Vaccine Wyeth Lederle 800.934.5556
* saccharides from capsular Streptococcus pneumoniae antigens (7 serotypes) individually conjugated to diphtheria CRM 197 protein aluminum phosphate, ammonium sulfate, soy protein, yeast

RabAvert - **Rabies** Chiron Behring GmbH & Company 510.655.8729
* fixed-virus strain, Flury LEP neomycin, chlortetracycline, and amphotericin B, potassium glutamate, and sucrose human albumin, bovine gelatin and serum "from source countries known to be free of bovine spongioform encephalopathy," and chicken protein

Rabies Vaccine Adsorbed GlaxoSmithKline 800.366.8900 x5231
*rabies virus adsorbed, beta-propiolactone, aluminum phosphate, thimerosal, and phenol, red rhesus monkey fetal lung cells

Recombivax - Recombinant **Hepatitis B** Merck & Co., Inc. 800.672.6372
* genetic sequence of the hepatitis B virus that codes for the surface antigen (HbSAg), cloned into GMO yeast, aluminum hydroxide, and thimerosal

RotaShield - Oral Tetravalent Rotavirus (recalled) Wyeth-Ayerst 800.934.5556
* 1 rhesus monkey rotavirus, 3 rhesus-human reassortant live viruses neomycin sulfate, amphotericin B potassium monophosphate, potassium diphosphate, sucrose, and monosodium glutamate (MSG) rhesus monkey fetal diploid cells, and bovine fetal serum smallpox (not licensed due to expiration)

40-yr old stuff "found" in Swiftwater, PA freezer Aventis Pasteur USA 800.VACCINE
* live vaccinia virus, with "some microbial contaminants," according to the Working Group on Civilian Biodefense polymyxcin B sulfate, streptomycin sulfate, chlortetracycline hydrochloride, and neomycin sulfate glycerin, and phenol -a compound obtained by distillation of coal tar vesicle fluid from calf skins

Smallpox (new, not licensed) Acambis, Inc. 617.494.1339 in partnership with Baxter BioScience
* highly-attenuated vaccinia virus, polymyxcin B sulfate, streptomycin sulfate, chlortetracycline hydrochloride, and neomycin sulfate glycerin, and phenol -a compound obtained by distillation of coal tar vesicle fluid from calf skins

TheraCys **BCG** (intravesicle -not licensed in US for tuberculosis) Aventis Pasteur USA 800.VACCINE
* live attenuated strain of Mycobacterium bovis monosodium glutamate (MSG), and polysorbate 80 (Tween-80)

Tripedia - **Diphtheria-Tetanus-Pertussis** Aventis Pasteur USA 800.VACCINE
*Corynebacterium diphtheriae and Clostridium tetani toxoids and acellular Bordetella pertussis adsorbed aluminum potassium sulfate, formaldehyde, thimerosal, and polysorbate 80 (Tween-80) gelatin, bovine extract

US-sourced Typhim Vi - **Typhoid** Aventis Pasteur USA SA 800.VACCINE
* cell surface Vi polysaccharide from Salmonella typhi Ty2 strain, aspartame, phenol, and polydimethylsiloxane (silicone)

Varivax - **Chickenpox** Merck & Co., Inc. 800.672.6372
* varicella live virus neomycin phosphate, sucrose, and monosodium glutamate (MSG) processed gelatin, fetal bovine

serum, guinea pig embryo cells, albumin from human blood, and human diploid cells from aborted fetal tissue

YF-VAX - **Yellow Fever** Aventis Pasteur USA 800.VACCINE
* 17D strain of yellow fever virus sorbitol chick embryo, and gelatin

http://www.informedchoice.info/cocktail.html
Vaccine Liberation Information
NOTE: THIMEROSAL = MERCURY
http://www.vaclib.org/pdf/exemption.htm

Vaccinations: Good, Bad or Just Plain Ugly

The FDA and other "watchdog" government agencies seldom are called to account for erroneous or irresponsible decisions. In the Dow Chemical silicone breast implant suit, the government was recently awarded $9.8 million for medical expenses paid out through Medicare and Medicaid. It didn't seem to matter that another agency, the FDA, of the same government had previously approved the use and sale of these implants and is currently considering whether to allow them to be sold again.

Further, these same agencies show definite bias when it comes to evaluating the risks associated with drugs. A good example is the fact that the agencies are constantly pushing for vaccinations and flu shots. For some reason, however, they neglect to tell the public that **the preservative in these flu shots and vaccines is mercury.**
        IS THERE SUCH A THING AS HEALTHY MERCURY?
When it comes to other sources of mercury, though, they are extremely vigilant. They have issued repeated warnings on the consumption of various fish, including tuna, shark, swordfish, and mahi-mahi, because of possible mercury contamination. And since mercury is particularly harmful to nerve cells, government health authorities have stressed that infants and small children shouldn't be fed these foods, and pregnant and nursing mothers should avoid eating tuna also.

However the facts state that most canned tuna contains less mercury contamination than tuna steaks, which come from larger tuna. It's hard to tell how much, if any, mercury these products contain. Smaller fish are safer, and so are fish like sole, sardines, herring, bass, catfish, salmon and shellfish.

Although the EPA (Environmental Protection Agency) has determined that the maximum allowable daily exposure to mercury is 0.1 microgram per kilogram of body weight, the new flu vaccine for babies, called Fluzone, contains 25 micrograms of mercury per 0.5 ml dose.

Practically all vaccines contain mercury and aluminum. And vaccines are not "safer" sources of these toxic minerals. It doesn't matter if the mercury comes from fish or from a vaccine. The potential for neurological damage remains the same. But for some reason, even though we're warned about fish consumption, vaccines and flu shots are strongly encouraged and, in many instances, even required by law. It shouldn't come as any surprise that more babies seem to be developing autism problems, and the risk of developing Alzheimer's disease is steadily increasing.

ALZHEIMER'S LINKED TO FLU SHOTS

In the year 2000, there were approximately 5 million people in the U.S. with Alzheimer's, and it has become the fourth leading cause of death in individuals over the age of 75. By the year 2010, it is estimated that over 7 million individuals will have the disease, and by 2025, 22 million will develop Alzheimer's.

As the general population continues to consume more contaminated food, water, and medicines, these predictions may very well prove accurate. One expert at the 1997 National Vaccine Information Center (NVIC) International Vaccine Conference stated that anyone who had five consecutive flu vaccine shots increased their risk of developing Alzheimer's disease by a factor of 10 over someone who received only two or fewer shots.

A powerful herb to prevent alzheimer's

It's worth mentioning, while we're on the Alzheimer's topic, that the elderly in India have the lowest incidence of Alzheimer's disease in the world. Only 1 percent of the elderly in India suffer from Alzheimer's. In contrast, the Alzheimer's Association in this country says that 10% of our population over 65 years old has the disease, and half of those over 85 have Alzheimer's. Researchers have theorized that the low incidence of Alzheimer's among the Indian population could be due to their increased consumption of the spice turmeric, a component of curry. Animal studies have supported this theory.

Studies have shown that when either turmeric or curcumin, (a major component of turmeric) was added to the diets of animals bred to develop

Alzheimer's, the brain damage was significantly lessened. [Neurobiol Aging 01;22(6):993-1005] [J Neurosci 01:21(21):8370-8377]

Turmeric has been shown to have very strong antioxidant properties that can be very effective at normal dietary doses. This spice may be one of the easiest and least expensive methods of combating the growing epidemic of Alzheimer's disease.

Better than a flu shot

When it comes to beating the flu, **selenium** can increase your odds. Selenium is a necessary mineral for the production of antioxidants within the body. New animal research from the University of North Carolina has found that a dietary deficiency of selenium may cause a harmless strain of the flu virus to mutate into a virulent pathogen.

When selenium-deficient mice were given a known flu virus and compared to mice with normal selenium levels, researchers found that the selenium-deficient animals experienced far more serious symptoms, such as lung damage. Based on this new research, other researchers are wondering if the more potent viruses, such as HIV, also mutated in environments where there were selenium deficiencies. It makes sense when you consider the well-known fact that most of the worldwide flu outbreaks originate in China, where large segments of the population are selenium-deficient.

Whether you decide to get flu shots or other vaccinations is a personal choice but as you weigh the pros and cons of such a decision, **don't be naive enough to think any of our government agencies have your best interests as their top priority. It could be a fatal mistake.**

THE VACCINATION

By Patricia Crutchfield

His trusting eyes looked up at me
He smiled his sweetest smile
What a precious gift from God he was
My son my first born child,

The nurse came in and weighed him
Put a thermometer briefly in his ear

Then she told me to take off his diaper
And expose his plump little rear.

I did as I was instructed
For I knew the procedure by now
It's time for his next vaccination
This time I won't flinch, I vow.

The syringes and vial of the serums
Lay benignly on her sterile steel tray
And though I try to watch her,
I find myself turning away.

His scream at the prick of the needle
Sends a bolt of pure terror through me
It's animal like pitch was not normal
And I turned around quickly to see.

His beautiful body went rigid
Then spasmed again and again
What's happening to my poor baby?
And what can I do to help him?

I could sense the nurse's pure panic
As she called out to the doctor to come
The seconds that passed seems like hours
And where is that screaming coming from?

I open my eyes in a room filled with light
The silence a deafening roar
My husband is standing beside me
He says everything fine, but his tears tell me more

I try to sit up, but I'm weary
Another needle pierces my arm
I drift off once again into darkness
But my mind beats a steady alarm.

Two days and two nights I am sedated
Until now no one tells me why
Then the doctor appears with my husband
And immediately I start to cry

My most precious gift has been taken
He'll never again be mine to hold
His body once so warm and loving
Now lays on a slab icy cold

I'm sorry says the good doctor
A reaction we couldn't foresee
Please accept my sincerest condolence
I guess it was just meant to be

Our son now plays with the angels
And my heart breaks anew everyday
Its the angels who tickle his tummy
And it's in their arms not mine, he will lay

A statistic, one in seventeen hundred
That's what they say of my son
But I say one child is too many
To die from a vaccination

So mothers do not be so trusting
Hear me before it's to late
Don't lose your child to the "program"
Investigate before you vaccinate
•••

Families Raise Concern Over Mercury In Vaccines
Debate Continues Over Past Use Of Thimerosal
POSTED: 1:37 p.m. EST November 4, 2002
 11:08 p.m. EST November 4, 2002 DURHAM COUNTY, N.C.

A growing contingent of parents believes a mercury-based preservative in those vaccines may have done more harm than good. In 1999, at the request of the Food and Drug Administration, drug companies agreed to begin removing a controversial preservative called thimerosal from vaccines. Some families believe the removal comes too late. Jackson Bono is a happy, curious 13-year-old challenged by a myriad of medical and developmental problems. Jackson has trouble speaking and focusing and works with a tutor.

"The toll it takes on a family is remarkable," said Scott Bono, Jackson's father. Like most parents, Scott and Laura Bono had their son vaccinated

when he was a baby. They now blame his problems on thimerosal and its main ingredient, mercury.

"Little did we ever suspect that the very immunizations that were to protect him from childhood diseases were poisoning him with mercury," Scott Bono said. Thimerosal kills harmful bacteria and has been in vaccines for decades. In the early 1990s, the number of recommended childhood vaccines increased. Over the last decade the national autism rate has risen drastically. In North Carolina, the rate has more than quadrupled, according to the state Department of Public Instruction.

Some people see a connection. If you add up the amount of mercury in baby vaccines with thimerosal, the levels exceed those considered safe for adults by the FDA. The Bonos said Jackson was a normal, healthy baby until he received a bundle of vaccines when he was 16 months old. They said, soon after, he stopped talking and making eye contact. Jackson developed autistic tendencies, like spinning uncontrollably. He also suffered severe allergies, seizures and stomach trouble.

"It was a cruel tragedy that happened with our son," Laura Bono said. Dr. Samuel Katz, chairman emeritus of paediatrics at Duke, is considered one of the foremost authorities on vaccines in the country. He raises doubts that thimerosal ever hurt children. "Whenever we have a problem, we like to know whose fault is it. Unfortunately, vaccines have become an easy target," he said. Katz said, "The evidence to support these claims is lacking." However, in 1999, he recommended drug companies take thimerosal out of vaccines. A 2001 report from the National Institute of Medicine also concluded the evidence does not support the claims. Researchers conceded, "the hypothesis is biologically plausible."

"Given that its mercury and we know that mercury has no beneficial effects, my statement to the FDA was that there's really no reason to use something like thimerosal," said Michael Aschner, a Wake Forest University neurobiologist. Aschner has studied mercury for 20 years. Research from the University of Calgary backs up his work and found mercury can destroy brain cells. Aschner points out that the ethylmercury in thimerosal is different from the damaging methylmercury found in some fish. He feels the issue clearly deserves much more study.

"If you do it in a dish, ethylmercury does cause significant effects, toxic effects. There's no question about it," Ascher said. "But, again, what you have to be careful of is how you translate what you see in a dish into a human being." The biggest obstacle parents of special needs children face in making the thimerosal argument is the fact that millions of children, a vast majority, got the same vaccine and never got sick.

"Why is it that all people who smoke don't get cancer? The body reacts differently to different antagonists," Salisbury attorney Bill Graham said. Graham represents 40 families who believe thimerosal hurt their children. He believes evidence is mounting that federal regulators knew that thimerosal could be harmful long before drug companies felt pressure to remove it from vaccines. A study sanctioned by the Centers for Disease Control and Prevention shows infants immunized with thimerosal vaccines were 2.5 times more likely to develop neurological disorders, but it was never released. Instead, the study continued and the results changed. Graham questions why vaccines were never recalled.

"Do you think that thimerosal vaccines that are potentially harmful could still be out there? They could be. They could be on the shelf right now," Graham said. "I really think the thimerosal issue has become a feeding frenzy. It's like the sharks with blood in the water," Katz said. The Bonos said they do not want blood. They want families like theirs to be heard for Jackson's sake, and others like him. "He's lost his childhood and he may not ever be what he should have been," Laura Bono said. Parents like the Bonos can file claims with the National Vaccine Injury Compensation Program. Because of the debate over thimerosal, the federal government has put all the claims on hold until further studies are completed. There was no recall of thimerosal vaccines, so it is possible some could still be on shelves. Anyone with concerns should talk to their child's pediatrician and ask for thimerosal-free vaccines. Both sides of the debate stress the importance of immunizing children.
Reporter: Cullen Browder
Photographer: Gil Hollingsworth
OnLine Producer: Michelle Singer

FOR VAST AMOUNTS OF INFORMATION CHECK:
http://www.vaccinetruth.org/

All truth goes through three stages. First it is ridiculed. Then it is violently opposed. Finally, it is accepted as self-evident."

(Schopenhauer)
****
"Condemnation without investigation is the height of ignorance." Albert Einstein
■■

"In the field of vaccination, medical training is simple indoctrination."
****
Inoculations are the true weapons of mass destruction causing an epidemic of GENOCIDE
Rebecca Carley, MD
Court Qualified Expert in VIDS (Vaccine Induced Diseases)
http://www.drcarley.com

..discussing vaccination with a doctor is like discussing vegetarianism with a butcher...........(George Bernard Shaw)

What good fortune for those in power that the people do not think."
~Adolf Hitler

"When you once see something as false which you have accepted as true, as natural, as human, then you can never go back to it" - J. Krishnamurti

It also gives us a very special, secret pleasure to see how unaware the people around us are of what is really happening to them."   ~Adolf Hitler

" Fear of disease, fear of microorganisms, fear of the unknown, is the tool of the clever that keeps the weak in line"   ~
Tim O'Shea, DC

What a strange religion medicine makes. It's the only religion that is federally backed, and even amid scientific controversy, cannot be questioned openly without persecution or ridicule."

Why Doctors do not understand the evils of vaccinations....
"It is difficult to get a man to understand something when his salary depends upon his not understanding it!"
Upton Sinclair

No one has ever successfully proven that any child has ever benefited from an injection of rotting matter combined with nerve and brain destroying poisons, the actual ingredients of vaccines. – Dewey

What is the name of the test that can be given to determine if a child can safely receive a vaccine?
It's called a breath test. You hold a mirror in front of the child and if condensation appears, they are still alive and cannot "safely" receive a vaccine. - steve

"Uneducated people believe what they are told...Educated people question what they are told"

"People do not like to think. If one thinks, one must reach conclusions. Conclusions are not always pleasant."
-Helen Keller

You can't wake a person who is pretending to be asleep. ~ Navajo Proverb
The art of medicine consists of amusing the patient while nature cures the disease—Voltaire

"A truth's initial commotion is directly proportional to how deeply the lie was believed...When a well-packaged web of lies has been sold gradually to the masses over generations, the truth will seem utterly preposterous and its speaker, a raving lunatic." --Dresden James

"He's the best physician that knows the worthlessness of most medicines."
"God heals and the Doctor takes the fee." - Benjamin Franklin, (1706-1790)

For us to bombard a newborn baby with a whole battery of vaccines as, in effect, their very first immunologic experience I think is reckless beyond measure. I would say it borders on the criminal.
Dr. Moscowitz

If you think that something is right just because everyone believes it, then you are not thinking" - Vievienne Westwood

Knowledge makes a man unfit to be a slave."
Frederick Douglass

Men occasionally stumble on the truth, but most of them pick themselves up and hurry off as if nothing had happened.
Winston Churchill

"First they ignore you, then they laugh at you, then they fight you, then you win." ~Ghandi

"Your failure to be informed does not make me a wacko." -- John Loeffler

I have no medical training, but I can read.

"The great tragedy of science - a beautiful hypothesis slain by an ugly fact." - Thomas Huxley

"If you think you're too small to be effective, you've never been in bed with a mosquito." - Betty Reese

"I know that most men, including those at ease with the problems of the greatest complexity, can seldom accept even the simplest and most obvious truth if it be such as would oblige them to admit the falsity of conclusions which they have delighted in explaining to colleagues, which they have proudly taught to others, and which they have woven, thread by thread, into the fabric of their lives."- Leo Tolstoy—

Right is right, even if everyone is against it; and wrong is wrong, even if everyone is for it --William Penn

Pediatricians Want to Keep Thimerosal in Vaccines Despite Health Risks

May 7 • Big Pharma, Health, Vaccines • 2039 Views • 0 Comments

Share484 0 Tweet0 Share0 1 Share488

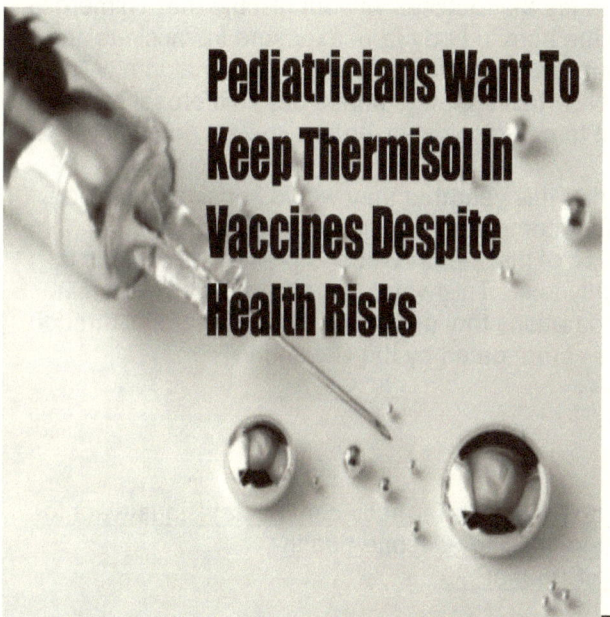

The American Academy of Pediatrics (APP) has released a statement in favor of Thimerosal, a mercury-based preservative that is detrimental to the health of everyone who is vaccinated.

The World Health Organization (WHO) and the APP are collaborating to say that Thimerosal should not be banned as an ingredient in vaccines in the US as well as a proposed banning of the toxin coming from the United Nations.

Dr. Louis Cooper from Columbia University states that research shows that Thimerosal does cause harm to children.

In 2004, the US Institute of Medicine conducted a safety review that found no evidence was viable to state that Thimerosal is connected to the soaring incidents of autism. The Centers for Disease Control and Prevention (CDC) concurred with this finding in a study released in 2010.

Thimerosal is ethyl mercury that has dramatic toxic effects against the health of the human body. The UN Environmental Program (UNEP) has listed mercury as a global health hazard. The UNEP would like to see a treaty that protects against Thimerosal be created.

The Coalition for Mercury-Free Drugs has provided the UNEP with researched evidence that outlines without dispute that Thimerosal is dangerous to the human body from exposure in vaccines and dental fillings that contain mercury-based silver amalgam. With this research, the UNEP has a plan to phase out Thimerosal because of its attachment to neurological disorders, autism and other health disorders.

WHO stated earlier this year that they would consider replacing Thimerosal, before completely changing their minds and supporting the use of this mercury-based preservative which is known to be highly toxic. This would save 84 million children in under-developed nations that are exposed to Thimerosal through vaccine initiatives propagated by the UN and WHO.

READ MORE
by Taboola
Popular Now

Read more at http://www.realfarmacy.com/pediatricians-want-to-keep-thimerosal-in-vaccines-despite-health-risks/#5scTSkKHFP69IIaF.99

THE FOLLOWING ARTICLE IS NOT A "CONSPIRACY THEORY", BUT A WELL-PUBLISHED AND REVEALED FACT. That will hurt some people's delicate mind who like to brand such as "Conspiracy th....."

How Pharmaceuticals Came To Be The 4th Leading Cause Of Death In America

Dec 1 • Articles, Big Pharma, The Word • 217 Views • 15 Comments

by LISA BLOOMQUIST

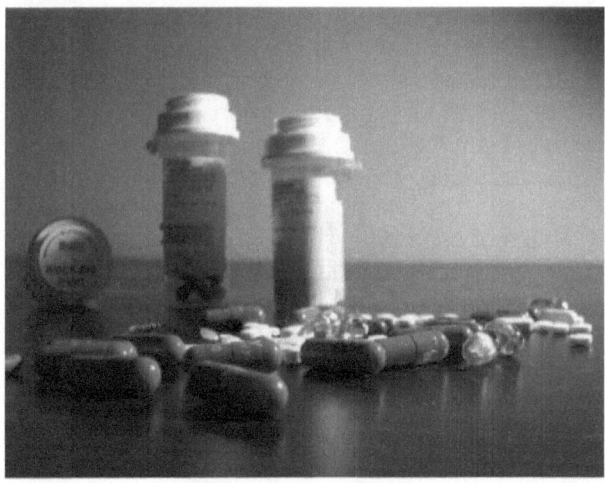

Prescription drugs are the 4th leading cause of death in America.

People know this to be true, they know it to be appalling, but it's still seen as incomprehensible and absurd. How could medicine hurt so many people? We all know that side-effects happen, but they are thought to be rare. They must be rare, right?

We all know some good, kind, generous, thoughtful doctors who want nothing more for their patients than health and happiness, so they certainly aren't giving their patients drugs that hurt them, are they? We know that the FDA is a federal bureaucracy, so it must be too restrictive of the pharmaceutical industry, right? And the FDA is supposed to protect consumers, so we're as safe as we can be, right? And people can sue, so the legal system must be keeping the bad aspects of the medical system in check, right?

All of these questions, and many more, **bring up some cognitive dissonance for people** when they're faced with the fact that prescription drugs, used as prescribed, kill an inordinate a number of people. It brings up the questions -

 How do prescription drugs get to be the 4th leading cause of death in America? How does that happen?

Here is a tale of how prescription drugs, used as prescribed, kill people.

Kerstin (age 30) comes down with a urinary tract infection. It's a Saturday so her regular doctor's office is closed. Urinary tract infections are painful so she knows that she can't wait 'til Monday to get treatment. She goes to an Emergency Services Clinic close to her house. She tells them that she has a urinary tract infection and they write her a prescription for Cipro (Ciprofloxacin – a fluoroquinolone antibiotic). They do not culture her urine because they don't have the time or capacity to do so. It doesn't matter what kind of bacteria is in her urine though, they know that Cipro will kill it because Cipro is a broad-spectrum antibiotic and it will kill all the offending bacteria in her urinary tract, plus some.

Kerstin is relieved. Her painful urinary tract and bladder are about to be healed.

 Kerstin takes two 500 mg. pills of Cipro two times a day for a week. On the 5th day of taking Cipro, Kerstin starts to feel a bit off. Her bladder feels full even when it isn't, she has dark "floaters" interfering with her vision and she feels anxious. She doesn't think anything of these things. They're strange, but not too worrisome. She doesn't think for a second that they could be due to the antibiotic that she is taking. Kerstin finishes the seven day course of Cipro. Her urinary tract infection is gone and she is pleased about that. Her bladder fullness, floaters and anxiety come and go and she doesn't think much of them.

Ten days pass in which Kerstin feels fine. On the eleventh day after she has finished taking Cipro, she starts taking ibuprofen to treat menstrual cramps. On the thirteenth day after she has finished taking the Cipro, it feels as if a bomb goes off in her body. Her hands and feet swell to twice their normal size. It becomes painful for her to walk or to do anything with her hands. Her knees are burning as if every tendon in them is inflamed. She is weak. She develops hives all over her body. Her anxiety levels are sky-high.

Kerstin goes to the doctor. The doctor says that the hives are a result of an allergic reaction and tells her to take Benadryl. Kerstin asks the doctor why she can barely walk when she was going to the gym daily just a few days earlier. The doctor says that she doesn't know, but that she will run tests.

Kerstin takes Benadryl but it doesn't seem to help. She goes back to the doctor for something stronger. She is put on prednisone.

The swelling in her hands and feet goes down, but her other symptoms worsen. She develops insomnia. She sprains her wrist while opening a jar. Intermittent pain throughout her body, but especially in her legs, begins. She loses her memory and has trouble concentrating.

Her test results come back. They are all "normal."

Her pain worsens. She is diagnosed with Fibromyalgia. She asks the doctor who diagnoses her with Fibromyalgia how she could have gone from being healthy and active to being disabled and in pain, now with a diagnosis of Fibromyalgia. The doctor mutters something about mysterious diseases and unexplained symptoms. Kerstin asks if her symptoms could be related to any of the medications that she took – Cipro, ibuprofen and prednisone. The doctor says no. More tests are run to see if there are other causes of Kerstin's symptoms.

Kerstin is put on Lyrica to help her with her Fibromyalgia pain.

The Lyrica seems to help some of her pain but her mental symptoms get worse while she is on it. In addition to her already existing memory and concentration problems, Kerstin develops brain-fog. She feels slow, stupid and like she is living in a dream. She gains 15 pounds in two months. Her hair starts to fall out. She feels suicidal. She is taken off of Lyrica by her doctor.

Kerstin continues to have problems in her joints, especially her wrists, knees and ankles, so she is not surprised when she is diagnosed with Rheumatoid Arthritis. She starts seeing a Rheumatologist who puts her on Humira. Humira decreases some of her inflammation symptoms but many of her other symptoms remain. She receives Humira treatments for 2 years.

After two years of Humira treatments, Kerstin is diagnosed with hepatosplenic T-cell lymphoma – cancer. She dies on the operating table when her surgeon attempts to remove the lymph nodes on her neck that had been affected by the cancer. She is 34 years old when she passes away.

Kerstin's story is fictionalized, but it is far from fantasy. Stories like hers happen every day. A large portion of her story is my own and it was both true and horrifying to experience. Stories that are significantly worse, where a doctor's injection site turn into a staph or MRSA infection to start the whole process, or where anti-psychotic medications that the patient are put on drive her to homicide or suicide. And I didn't delve into the PAIN that comes with Fluoroquinolone Toxicity (Cipro is a fluoroquinolone and the others are just as bad, if not worse), so it's a light fictionalized version – with the hope that you'll find the horror to be believable, because it is very, very real for too many people.

The Explanations, Journal Articles and Facts behind Kerstin's Story
I don't expect you to accept the story above as fact without some thorough explanation. Here is the information behind my assertions:

The antibiotic that Kerstin took is Cipro (Ciprofloxacin). Cipro is a fluoroquinolone antibiotic, along with Levaquin, Avelox, Floxin and a few other less commonly used drugs in the fluoroquinolone class. Fluoroquinolone antibiotics are the "big guns" of antibiotics.

They are broad spectrum antibiotics that will kill all bacteria in their path. (2) They are frequently prescribed to treat urinary tract infections (3) because they penetrate kidney tissue well (4).

Cipro, and all the other fluroquinolones, have terrible side-effects. Many of the awful side-effects that can be experienced, often all at once, are listed on the Cipro Warning Label. However, many things are left off of the warning label, they are listed on http://www.ciproispoison.com/.

Additionally, here are articles backing up Kerstin's symptoms:

- Vision Floaters – The JAMA article entitled "Oral Fluoroquinolones and the Risk of Retinal Detachment" notes that fluoroquinolones increase the incidence of Retinal Detachment (5). If the connective tissue in your eye is damaged, visual disturbances, including floaters, can result.

 Anxiety – The Journal of Neurosciences in Rural Practices' article entitled "Levofloacin-induced Acute Anxiety and Insomnia" notes that Levofloxacin (another fluoroquinolone – Levaquin) can induce anxiety and insomnia (6) Cipro/Ciprofloxacin can do the same.

- Bladder fullness – This is a symptom that I experience, along with many other people suffering from Fluoroquinolone Toxicity. I'm not completely sure what it stems from, but here are a couple of possibilities. This article in the Journal of Urology entitled "Role of Mitochondria in Ciprofloxacin Induced Apoptosis in Bladder Cancer Cells" notes that Cipro disturbs the mitochondria in bladder cells and causes apoptosis (cell death) (7). It is also possible that the feeling of bladder fullness is a result of dysglycemia as it is noted in an article in Medscape Medical News that fluoroquinolones increase the risk of severe dysglycemia in diabetics. Additionally, "one fluoroquinolone antibiotic, gatifloxacin (*Tequin*, Bristol-Myers Squibb), was already withdrawn from the US market in 2006 due to the risk for severe dysglycemia" (8)

- Pain and swelling in hands and feet – This symptom can be more succinctly described as peripheral neuropathy. The

FDA issued an update to the labels for fluoroquinolones noting that PERMANENT Peripheral Neuropathy is a possibility in August, 2013 (9). This neuropathy may stem from destruction of the Myelin caused by the fluoroquinolone.

There are likely other causes and reasons for Peripheral Neuropathy being a result of Fluoroquinolone Toxicity, including the production of neurotoxins caused by the drugs (10) and the fracturing of DNA (11).

- Skin problems like hives/uticaria/rashes are listed on the warning label

- Tendon pain/tear/strain/rupture – This adverse effect is so well documented that fluoroquinolones carry a black box warning about the danger of rupturing a tendon on the top of the warning label. An article in Musculoskeletal Medicine entitled "Musculoskeltal Complications of Fluoroquinolones: Guidelines and Precautions for Usage in the Athletic Population" notes that young, healthy, athletic people's muscles and tendons are adversely effected by fluoroquinolones (12)

- Fibromyalgia – Mysterious, sometimes intermittent, sometimes constant, pain is common among those with fluoroquinolone toxicity. The information above about peripheral neuropathy should ring a lot of bells for those diagnosed with Fibromyalgia. Additionally, Carboxylic Acid is attached to the quinolone molecule (13). It is a known neurotoxin. (14 and 15) Also, a quinolone studied in the article "Cytotoxcicity of Quinolones toward Eukaryotic Cells" notes that quinolones "kills cells by converting the (topoisomerase) type II ezyme into a cellular poison." (16) Cellular poisons can lead to pain.

- A diagnosis of Rheumatoid Arthritis – Per Toxicologist, Professor Joe King, "when a cell is malfunctioning (due to a mutation caused by a toxin or radiation) the body deems it alien and begins and autoimmune response as a defense mechanism. Thus producing positive autoimmune antibodies in lab tests, but in actuality you don't really have the disease, it is bad cells. For example I test positive for rheumatoid arthritis (RA), but I don't have RA, I have Fibrillan Connective Tissue destruction upon biopsy. But the doctor reads the lab report for RA and recommends anti-inflammatory steroids. **Bad diagnosis**, because the problem is not RA but Fibrillan and steroids will dissolve the Fibrillan faster." Also per Professor King, "the cells in your tissue, organs, etc. are not functioning correctly, there is a mutation in there somewhere and the body is reacting to this weird cells as alien, thus producing an inflammatory process (which is painful)." Additionally, it should be noted that Cipro was found to cause chromosomal abnormalities in immune system cells. (17)

I mentioned NSAIDs and steroids. Both NSAIDs and steroids are contraindicated with fluoroquinolones (18 and 12). Please note that Kerstin didn't take NSAIDs or steroids at the same time as the Cipro. Both NSAIDs and steroids are contraindicated for any person who has ever experienced an adverse reaction to a fluoroquinolone, likely because of the production of acyl glucuronides, "which are chemically reactive electrophiles formed by carboxylic acid-containing drugs" (15) and/or because of the depletion of the CPY450 enzymes by quinolones/fluoroquinolones that leave the body unable to metabolize other drugs (19 and 20).

How do fluoroquinolone ANTIBIOTICS cause all that harm? The harm that they cause is in the essence of the way they work. They are the "first antibacterial agents that efficiently inhibited DNA replication." (21) Antibiotics in the penicillin and cephalosporin classes, by comparison, work by disrupting bacterial cell walls, not by doing anything to bacterial, or human, DNA. Fluoroquinolones also form "a poisonous adduct on DNA" (21). Fluoroquinolones cause chromosomal abnormalities in human cells (17) and also have tumor killing qualities (22). While that might sound great on the surface, if you read between the lines you'll note that if these drugs kill tumor cells, they kill human cells. Fluoroquinolones cause apoptosis, programmed cell death, at a massive rate (23). Patient studies have shown, through a DNA Adduct Mass

Spectrogram Analysis, that quinolone molecules have adducted to their DNA. Adducting to and breaking human DNA can cause every single one of the problems that Kerstin experienced, all of the problems listed on the FDA warning label for these drugs, and more. It's a bad idea to mess with human DNA and chromosomes – a look back at the history of Agent Orange will tell you why this is true.

The consequences of the DNA destruction done by fluoroquinolones is yet to be established. An article was published in Nature in September, 2013 connecting topoisomerase inhibiting drugs (fluoroquinolones inhibit topoisomerases II and IV (24)) with triggering the expression of autism related genes. I wrote about this on CE – http://www.collective-evolution.com/2013/09/18/a-horrifying-cause-of-autism-dna-damage-from-synthetic-antibiotics/ Of course, more studies need to be done to determine the implications of this study.

Studies of the DNA make-up of Gulf War Veterans and their children may also be revealing as all 1991 Gulf War Veterans were given Cipro prophylactically because of fear of anthrax (25). Likewise, in 2001 United States Postal Workers who took Cipro prophylactically, also to prevent anthrax, and any ensuing health issues that they have (57% reported side-effects -26) may be related to their exposure to fluoroquinolones.

Fluoroquinolone antibiotics are dangerous drugs that have been used recklessly since their introduction to the market as a first-choice broad-spectrum antibiotic. They are likely responsible for many of the "mysterious" illnesses that have been on the rise since the early 1980s when Cipro was patented by Bayer and Levaquin was patented by Johnson & Johnson. Everyone who has Fibromyalgia, Chronic Fatigue Syndrome, Thyroid Dysfunction, any Autoimmune Disease, Gulf War Syndrome, Leaky Gut Syndrome, Dysautonomia, etc. should look at their medical records to see if they have ever taken a fluoroquinolone. If a fluoroquinolone is in your past, fractured genes may have resulted, and thus your pain and suffering. Please note that adverse reactions to fluoroquinolones are often delayed for weeks or sometimes months or years after administration of the drugs has stopped and there is a tolerance threshold for metabolism of these drugs (20) so most people do not react to their first dose.

Lyrica and Humira

Here is the warning label for Lyrica – (link – Source 27) Please note that suicidal ideation is one of the acknowledged adverse effects caused by this drug. Weight gain, difficulty concentrating, etc. are also listed on the warning label. Patient reports (these people aren't lying) can be found on askapatient.com – Lyrica.

Humira, Enbrel and other TNF inhibiting drugs CAUSE CANCER. This is well documented and known. The warning labels for both Humira and Enbrel state in a big black box warning that various cancers are associated with use of those drugs. In case it needs to be spelled out, cancer can be deadly.

Here is an excellent blog post about how Humira can kill, and how it is marketed – http://davidhealy.org/welcome-to-the-humiraverse/

Conclusions

It is often noted as people are bemoaning the unwillingness of the pharmaceutical industry to create more antibiotics, that there isn't enough money to be made from antibiotics to encourage their production. (28) While there may not be much money to be made in selling antibiotics directly, there is a whole lot of money to be made in treating autoimmune diseases. Humira reached $7.9 Billion in sales (29) in 2011 despite the undisputed fact that it causes cancer. If a class of antibiotics can cause the body to react as it would if it had an autoimmune disease for an extended period of time (the ill effects of fluorouquinolones can be permanent but they typically last from 6 to 36 months), and therefore a person gets diagnosed and treated for an autoimmune disease, though they don't actually have the autoimmune disease, they actually have an autoimmune reaction to a poisonous drug, the pharmaceutical industry has effectively taken an acute problem, an infection, and converted it into a chronic problem, an autoimmune disease. Chronic conditions mean repeat customers and the pharmaceutical industry makes billions. (I doubt that this process is a conspiracy or even planned on the parts of the people in the pharmaceutical industry. Rather, I think that it is caused by willful ignorance among those in the medical professions, encouraged by greed and a complete lack of checks and balances on the pharmaceutical companies, those that have the most to gain in creating repeat customers.)

People are being hurt by their medicine and it is unacceptable. If harm is impossible to avoid completely, it should be minimized. There is zero effort on the part of Doctors, Pharmacists, the FDA or anyone else to minimize adverse effects of drugs. If an effort were being made, we would not be in the tragic situation that we're in, with the pharmaceutical industry being the 4th leading cause of death of Americans.

The mantra of "all drugs have side-effects" has been so ingrained into the collective consciousness that we have come to think of it as acceptable that drugs have side-effects, and for drug side-effects to be devastating. In accepting this "better someone else than me" / "it can't happen to me" attitude, we have given permission to the FDA to be inept, incompetent and lazy. In their ineptitude, they have ignored 15 years of research noting that commonly prescribed ANTIBIOTICS are damaging our DNA. We can only hope that this oversight caused by laziness and incompetence is not consequential to us all. Because I can accept the possibility that it may be worth it for society for me to be sacrificed so that we can have powerful antibiotics, but no drug of any sort, no matter what good it does, is worth sacrificing our collective DNA.

Read more at http://www.realfarmacy.com/how-pharmaceuticals-came-to-be-the-4th-leading-cause-of-death-in-america/#7oSAWuxG4yDElvpM.99

Facts About Fluoridation – New Zealand

Douglas Main, OurAmazingPlanet Staff Writer | June 03, 2013 07:10pm ET

What is fluoridation?

Fluoride is an ionic compound containing <u>fluorine, which is the single most reactive element</u>; it is naturally found in many rocks.

Fluoride is added to public water supplies at an average concentration of about 1 part per million (1 ppm), or slightly below. Naturally occurring fluoride concentrations in surface waters depend on location but are generally low and <u>usually do not exceed 0.3 ppm</u>. Groundwater can contain much higher levels, however.

Exactly <u>how fluoride helps protect teeth, and how much it protects them, however, isn't completely clear.</u>

Tooth decay has declined in the United States since fluoridation began; **however, it has also declined in other countries that do not fluoridate.**

Those on the opposite side say that **it is unethical form of mass-medication, without each individual's consent or knowledge.** By putting fluoride in drinking water, the dosage cannot be controlled, since some people — like laborers and people with kidney problems — drink much more water than others. Fluoride opponents cite studies showing that low levels of fluoride have been linked to a number of negative health effects like bone fractures, thyroid disorders and impaired brain development and function.

The most obvious health effect of excess fluoride exposure is dental fluorosis, which when mild includes white streaks, and when severe can include brown stains, pits and broken enamel. **As of 2010, 41 percent of kids ages 12 to 15 had some form of dental fluorosis, <u>according to the CDC</u>.**

Within the last 15 years, research has revealed that fluoride primarily works topically, such as when it applied to the teeth in <u>fluoride-rich toothpaste</u>. People opposed to fluoridation have argued out that since this is true, it needn't be added to water. **Today people are exposed to many more sources of fluoride** than when it was first introduced in the 1940s — the first fluoridated toothpaste, Crest, wasn't introduced until 1955.

A 2009 study that tracked fluoride exposure in more than 600 children in Iowa found no significant link between fluoride exposure and tooth decay. Another 2007 review in the British Medical Journal stated that "there have been no randomized trials of water fluoridation," which is currently standard for all drugs.

Is fluoride bad for you? "Scientific" research confirms it is damaging and causes I.Q. damage

It depends who you ask; **fluoride is unquestionably toxic** at certain concentrations.

One study published in the fall of 2012 in the journal Environmental Health Perspectives found a link between high fluoride levels found naturally in drinking water in China and elsewhere in the world, and lower IQs in children. The paper looked at the results of 27 different studies, 26 of which found a link between high-fluoride drinking water and lower IQ. The average IQ difference between high and low fluoride areas was 7 points, the study found.

Finding the truth about fluoride

By Geoff Cumming

No wonder the forces of control are losing the battle to drug   the water supply with toxic scrapings   from industrial chimneys - acids which   the "antis" link to lower IQ, brittle bones, thyroid problems, male infertility, cancer and   Adolf Hitler. (Go online and grow very afraid, but then check the veracity and depth of   the research.)

Dismissed for so long as fringe nutters and conspiracy theorists, anti-fluoride campaigners keep bouncing back and, finally, seem to be winning. First they took Ashburton. In the current (2010-13) local government term, Taumarunui, New Plymouth, Central Hawkes Bay and now Hamilton have fallen - the last three after comprehensive "tribunal" inquiries.

Suddenly, less than half the population lives in areas with fluoride-added water and there's little impetus in non-fluoridated areas to reverse the tide.

Hastings and Whakatane loom as the next battlefields with referendums in the coming local body elections. Root work, in the form of protests and public meetings, is under way in Auckland and Wellington to put fluoridation on the agenda in next year's council annual plan processes. Could the silent majority, who've accepted the assurances of the medical establishment for so long, now be a minority? <u>Were public health authorities wrong about one of the biggest mass medication programmes in modern times?</u>

What's beyond challenge is that the anti-fluoride lobby, though small and shoestring-funded, has become well-organised and highly-skilled at challenging the establishment. The internet has helped local campaigners to access international resources, raise funds and run campaigns. Community groups can tap into a national body with links to the US-based Fluoride Action Network. They are no longer isolated but <u>part of an enlightened global movement. For every comprehensive, peer-reviewed study which supports fluoridation, there are many more studies raising doubts about safety and efficacy</u>, no matter how reliable or relevant.

Local Government NZ wants the issue taken away from councils and decided at central government level. Hamilton Mayor Julie Hardaker, a lawyer who says she came new to the issue after winning the mayoralty with no council experience, agrees. "Measures to address tooth decay are a public health matter and in this country public health is dealt with by central government. <u>We deal with delivering the water and purifying it to drinking water standards</u> - we don't deal with medical issues."

Many councillors support Hardaker's view that fluoridation raises complex scientific questions which politicians should not have to call. But the Government isn't about to risk fallout by taking the decision away from communities.

"The Government's role is to support local councils when they decide to use fluoride - not to make the decision for them, and we are not considering changing that," Health

Minister Tony Ryall told the *Weekend Herald.* **After New Plymouth opted out in 2011, Labour promised to hold an independent inquiry (the last was in 2001) with a view to developing a national policy when next in power but the Greens at the time expressed reservations about "mass medication of water" and support for local decision-making.**

Responsibility for defending (and promoting) fluoridation rests with district health boards - and their public health staff generally have more pressing priorities such as rheumatic fever or meningitis outbreaks to attend to. Repelling the antis' arguments is both time-consuming and thankless.

Where the antis have gained most traction is in tribunal-style hearings, a process adopted by the New Plymouth, Central Hawkes Bay and Hamilton councils to decide whether to retain fluoridation. For councils, this quasi-court process holds the promise of getting a perennial issue off the agenda: a decision based on an open, exhaustive inquiry shouldn't have to be revisited in a hurry.

Hardaker says Hamilton opted for a tribunal in preference to yet another referendum (a 2006 vote found 70 per cent support for fluoridation but was challenged as unfair), largely because all sides felt the tribunal process was fair. The Hamilton hearing took four days - time enough, outsiders might think, for medical experts to debunk the myths, quell the emotion, and convince politicians of the wisdom of a measure supported by the World Health Organisation and key health agencies in the United States, Britain, Australia and New Zealand. Instead, Hamilton councillors voted 7 to 1 to end fluoridation. It may not have been that clearcut: three councillors, on legal advice, withdrew from the process because they are elected members of the Waikato District Health Board. Two others stood back because they had expressed support for fluoridation and could be accused of bias. But even if the five had voted for the status quo, they might still have lost.

What went on?

DENTAL "INDUSTRY" PUSH FLUORIDE.

The district health board gave it its best shot, putting up the chief dental officer, medical officer of health, Maori dental health experts

backed by Ministry of Health's chief child health adviser, chief dental officer and a legal adviser.

But the Fluoride Action Network brought some highly-credentialled experts of its own: doctors and dentists including Dr Andrew Harms, former president of the South Australian branch of the Australian Dental Association; Hamilton Accident and Emergency doctor Peter Scanlon; Whangarei dentist Lawrie Brett and oncologist Anna Goodwin.

Experts who've turned - who once stood for fluoridation but now oppose it, such as former Auckland principal dental officer John Colquhoon - helped to raise doubts over the certainty of experts. Video links to strident international campaigners Paul Connett (USA) and Declan Waugh (Ireland) further muddied the scientific waters. (Connett, of the Fluoride Action Network, has made three New Zealand tours since 2003 and, wherever he goes, fluoridation revolts tend to follow).

Mary Byrne, national co-ordinator of the Fluoride Action Network NZ, says campaigners have always included medical professionals but as councils have grown more open-minded it's become easier to go public. "Having doctors and dentists and scientists gives [decision-makers] that level of security - that it's not just people misinterpreting science or peddling pseudo-science."

The net effect, according to councillors, was that the experts - along with the research evidence - cancelled each other out; their decision to stop fluoridation swung on issues other than science.

Several councillors were influenced by the mathematical blurring of individual vs population gains and bought the anti-fluoridists' line that "at worst it's equivalent to one further decayed tooth over your lifetime".

"One other thing was a real arrogance from some of the DHB people and scientists who told us we'd just been listening to a whole lot of quack science," says councillor Dave MacPherson. "We felt they were treating us as a bunch of morons." (& ISN'T THAT EXACTLY HOW THEY ACT. Bob.)

To emphasise what they are up against, the district health board used a shocking image of a child's front teeth rotted by decay in a non-fluoridated area. But fluoridation opponents used the same image to argue the real culprit was sugar.

Evidence that fluoridated water plays a minor role compared to topical application of fluoride toothpaste swayed some, particularly those attuned to <u>concepts of individual responsibility - and choice</u>.

"For people who don't want fluoride in the water, it's extremely costly to remove it [using filters]," says Hardaker, "whereas people who want to use fluoride can buy a three-pack of toothbrushes for 99c at Pak'nSave and a tube of toothpaste on special for $1.29 - and the best application is to clean your teeth twice a day."

<u>"The other issue was babies," Hardaker says. "Advice from the Ministry of Health was that children under 3 should not ingest fluoride and that babies under 6 months should not be exposed to fluoride via bottle-fed formula."</u> (But the 'authorities' put it in water for 100% of people, babies, invalids, pets, even car washes and laundry etc. Bob.)

There's nothing new in the arguments raised by opponents - their strategies and the studies they draw on have been critically analysed by Adelaide-based fluoridation supporter Jason Armfield of the Australian Research Centre for Population Oral Health. "Techniques such as the 'big lie' and innuendo are used to associate water fluoridation with health and environmental disasters," Armfield wrote in a 2007 analysis. "Half-truths are presented, fallacious statements reiterated and attempts are made to bamboozle the public ..." (But does this idiot give any useful information to evidence they are genuinely in error? No, he calls their words "innuendo" and half truth, without unfailingly evidences such. Bob)

It's clear the two camps are deeply locked in trench warfare; not just at referendum or tribunal time.

It can go too far - Waikato DHB boss Craig Climo complains that senior officials have been subjected to offensive, highly-personal emails before and since the Hamilton result. (moan moan)

It's relevant that council decision-makers, not just in Hamilton, are predominantly white and middle class, while the majority of their constituents behave as if there really is something in the water - improved dental health has fuelled complacency over fluoridation.

What irks authorities most is that the measure continues to have value as an ambulance at the bottom of the cliff - giving socio-economically deprived groups, particularly Maori and Pacific Island children, a fighting chance of healthy teeth and adults a better chance of keeping theirs. (This is bogus innuendo indeed. Bob)

These groups are more likely to have higher intakes of high-sugar drinks and snacks, and less likely to brush their teeth as frequently. Many under-5s don't access any dental care, Waikato DHB principal dental officer Rob Aitken told the tribunal. (The point is? This is surely hardly an issue supporting fluoridation in a case that "MAY" improve BAD diet etc of a minority, by imposing compulsory medication to the 100% population including babies, as mentioned above. Easier to educate that minority in dental self-card. Scarcely remotely a valid point of support to fluoridation. Bob.)

A truncated trial in Northland points to the difference fluoridation can make in disadvantaged communities. During the two-year trial, decay rates fell significantly in newly fluoridated areas in Kaitaia and Kaikohe, even though advocates thought the benefits might need five years to show up. But the Far North District Council ended the trial in 2009 after a postal vote which drew a response of just 16 per cent.

Waikato DHB estimates the end to fluoridation in Hamilton will add $500,000 a year to the cost of treating decay. (Yeah right, scare scare.)

Ministry chief dental officer Dr Robyn Haisman-Welch stresses that the fluoridation tide has not all been in one direction. Thames, Dunedin, Hutt Valley and Kapiti Coast all recently resolved to maintain treatment after passionate debates. (It would seem obvious that the fluoride works well as intended in those areas. Bob)

Treatment has been introduced in Patea and Waverley in Taranaki while the Waikato District Council is working to extend it to northern Waikato towns. The 2009 oral health survey showed public support for fluoridation was still strong, Haisman-Welch says. (Such comments as A or B did such and such is NOT the issue. If such and such are fat, should all become fat? This is the same spurious talking. Bob)

But the pro-fluoride camp needs to lift its game. If New Plymouth wasn't enough of a wake-up call, Hamilton certainly was, with the

Prime Minister's chief science adviser, Sir Peter Gluckman, issuing a missive and the ministry reviewing its approach. (It soon becomes obvious on whose "side" the writer of the article supports, n'est pas? Bob)

A ministry offshoot, the National Fluoridation Information Service, keeps tabs on the fluoride science industry (such is the volume of research) and <u>maintains that none of the health fears raised by opponents amount to a proven threat</u> - not in the dosages allowed in our water of less than 1 part per million. (There goes the famous denial words yet again by "Authority" <u>**not a "proven threat".**</u> No, they're not looking and never will look and see. Bob)

That doesn't mean there isn't room for doubt. <u>The one established side effect is dental fluorosis,</u> a mottling or flecking of tooth enamel. This has an aesthetic effect but is not a clinical concern at current levels. Children under 6 should not use full-strength toothpastes and there is concern about the impact of formula milk on babies' developing teeth.

But Emmeline Haymes, information service national co-ordinator, suggests more is needed than reassuring messages. DHBs do not have staff dedicated to the issue and do not take a pro-active approach. "Other than me, there's nobody working permanently on this in the country."

By <u>Geoff Cumming</u> <u>Email Geoff</u>

<u>An overwhelming majority of the more than 400 submissions received by council on the fluoridation issue were against the practice.</u>

WHAT IT MEANS

All fluoridation will cease in the New Plymouth district within approximately six weeks. Fluoride has no taste or smell.

Removing it will not change the taste of drinking water. The New Plymouth water supply has been fluoridated since 1969, Waitara and Lepperton since 1990, and Urenui since 1999.

The Inglewood, Oakura, and Okato water supplies are not fluoridated.

111 litres of fluoride is added to the New Plymouth district water supply every day. Water is fluoridated with hydro-fluorosilicic acid (HFA). The New Plymouth District Council spends $18,000 each year to fluoridate water.

The Ministry of Health recommends fluoridating water supplies to reduce dental decay. The level of fluoride naturally occurring in New Plymouth water is: 0.05g/m3. After treatment: 0.7g/m3. About 2.3 million people are supplied with fluoridated water in New Zealand.

- © Fairfax NZ News

Wellington region costs etc

What type of fluoride do we use?

Sodium silicofluoride (SSF), a powder, is used at our Te Marua, Wainuiomata and Waterloo treatment plants. The SSF we use has a minimum purity of 98.5% and is regularly tested to ensure that the minute quantities of other chemical elements, such as metals, that it contains are well within the maximum safe limits described in the Drinking Water Standards for New Zealand.

Hydrofluorosilicic acid (HFA), a liquid, is used at our Gear Island treatment plant. HFA has a fluoride content of not less than 15% and is typically around 18%. As for SSF, the HFA we use is tested to ensure that any other chemical elements of health significance that it contains are at safe levels in relation to the Drinking Water Standards for New Zealand.

What does water fluoridation cost?

The total annual cost of fluoridating the water that we supply is approximately $220,000 or around 55 cents per head of population supplied.

Fluoride will no longer be added to drinking supplies in New Zealand's fourth-largest city, ending nearly five decades of established practice.

Councillors voted this morning to stop adding fluoride to Hamilton's drinking water, overturning the outcome of a 2006 referendum in support of continued fluoridation.

The decision will create the second-largest population in the country without access to fluoridated water (after Christchurch). Hamilton joins a growing number of smaller communities that have voted to reduce or eliminate fluoridation, including New Plymouth, Central Hawkes Bay, Dunedin and Tauranga.

DENTAL INDUSTRY FILIBUSTERING

But can we really trust them with their past record?

Dr Jonathan Broadbent, Senior Lecturer in Preventive and Restorative Dentistry, University of Otago and President, Otago Branch of NZ Dental Association, comments:

"This decision will undermine public health in New Zealand. We are going in the wrong direction. Right now our nearest neighbors, the Australian State of Victoria, are spending Au$3.6 million build more fluoridation plants across rural parts of their State.

*"The World Health Organisation, **the World Dental Federation, and the International Association for Dental Research** have all stated that 'universal access to **fluoride for dental health is part of the basic human right to health'.** In New Zealand, a central part of the universal right to fluoride is community water fluoridation. The New Zealand Ministry of Health Guidelines and Statements (2010) on fluoridation are clear: community water fluoridation is effective and safe, and community water supplies in New Zealand should be fluoridated at 0.7-1.0 parts per million (ppm) wherever feasible. **The 7 Councillors who voted against this in Hamilton were either unaware of this, or disregarded it** (as well as disregarding the opinion of their own citizens from a 2006 referendum).*

"Those who are unwilling to drink fluoridated water should not be permitted to impose the risks, damage, and costs of failure to fluoridate on others. The ethics and science in support of fluoridation are clear, but antifluoridation arguments often present a highly misleading picture of water fluoridation.

"While the extent of tooth decay has reduced in recent decades, the disease remains more prevalent than other significant health conditions in New Zealand (such as asthma), particularly in unfluoridated areas and among disadvantaged New Zealanders. The recent New Zealand Oral Health Survey found much less tooth decay in fluoridated than non-

fluoridated areas. There is generally 0.3 ppm background fluoride in New Zealand (although it varies), and adjusting that to Ministry of Health-recommended levels has a significant effect of reducing tooth decay among people of all ages.

"There is a very strong public health case for expanding the use of fluoridation. To a large extent, the improved oral health we now enjoy in New Zealand compared to the past is due to water fluoridation. We forget that very few New Zealanders are affected by water-borne diseases, thanks to water chlorination. If chlorination was not used, we would have more water-borne diseases. Fluoride does the same for tooth decay.

"While community water fluoridation is known to be effective and safe, Hamilton is the fourth town in the past 2 years to remove it. It is time for the Ministry of Health and central government politicians to properly support the policy of community water fluoridation and provide a cohesive public health policy for universal access to fluoride in our communities, and legislate to fluoridate. We have very little in the way of a 'safety net' for adults who cannot afford dental care. Now that water fluoridation in Hamilton is being removed, we can expect dental caries experience in that city will increase, and health inequalities will increase too. Following this decision, I hope that Hamilton can put a plan in place for the dental care of those that cannot afford it."

What's in your water?

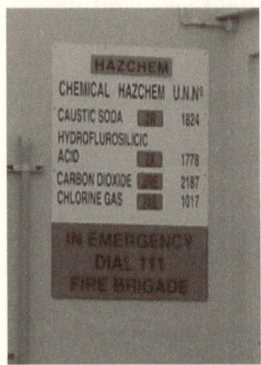

Hydrofluosilicic acid: S7 (Dangerous Poison)

This material is hazardous according to criteria of NOHSC; HAZARDOUS SUBSTANCE.

Subclasses:
6.1 Category D: Substances which are acutely toxic.
8.1 Category A: Substances that are corrosive to metals.
8.2 Category B: Substances that are corrosive to dermal tissue
8.3 Category A: Substances that are corrosive to ocular tissue

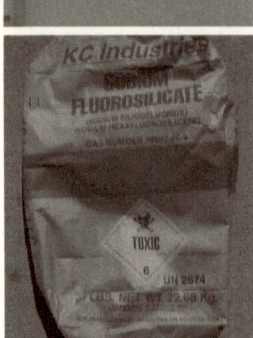

Sodium fluorosilicate: S6 (Poison)

This material is hazardous according to criteria of ASCC; HAZARDOUS SUBSTANCE.
Subclasses:
6.1 Category C: Substances which are acutely toxic.
6.4 Category A: Substances that are irritating to the eye.
9.3 Category B: Substances that are ecotoxic to terrestrial ver

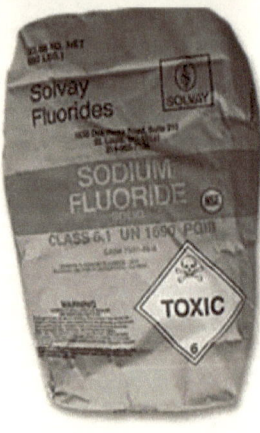

Sodium fluoride: S6 (Poison)

This material is hazardous according to criteria of Safe Work HAZARDOUS SUBSTANCE.

Subclasses:
6.1 Category C: Substances which are acutely toxic.
6.3 Category A: Substances that are irritating to the skin.
6.4 Category A: Substances that are irritating to the eye.
6.6B Category III: Substances which cause concern for huma possible mutagenic effects.
6.8 Category B: Substances that are suspected human reprodu developmental toxicants.
6.9 Category A: Substances that are toxic to human target org systems.
9.1 Category D: Substances that are slightly harmful to the aq environment or are otherwise designed for biocidal action.
9.3 Category B: Substances that are ecotoxic to terrestrial ver

These substances have never been approved for human ingestion, or studied for long term health effects. Yet, they are what is being added to our municipal water to provide fluoride. In New Zealand, all District Health Boards have been instructed to pressure local councils to fluoridate any unfluoridated water supplies, regardless of the wishes of individual

"I am appalled at the prospect of using water as a vehicle for drugs. Fluoride is a corrosive poison that will produce serious effects on a long range basis. Any attempt to use water this way is deplorable."
- *Dr. Charles Gordon Heyd, Past President of the American Medical Association.*
"Fluorides are general protoplasmic poisons, with the capacity to modify cell metabolism, changing the permeability of the cell membrane by inhibiting certain enzymes. Sources of fluoride intoxication include drinking water containing 1ppm or more of fluorine."
-*Journal of the American Medical Association, September 18, 1943.*
How did it suddenly become "safe"?

"Fluoridation is against all principles (modern pharmacology. It's obsolete. I think a single dentist would bring up t question in Sweden anymore."
- *Dr. Arvid Carlsson, co-winner of the Nobel Prize for Medicine (2000)*
"The mutagenic properties of fluoride: have been demonstrated by experimer carried out under the strictest scientifi conditions. "
- *Quebec Ministerial Inquiry into Fluoridation 1979.*
"..we now know that fluoride doesn't to be swallowed."
- *Dr Hardy Limeback, former head of Preventative Dentistry, Toronto Unive*
"Fluoridation is a naive utopia, withou practical effect, and an attack on perso liberty."
- *Minister of Health, Luxembourg, 19*

Photo: fountain in Petone that provides pure, untreated, naturally clean artesian water. Hundreds of people come every week from all around Wellington to fill up their water supply.

www.ingramcontent.com/pod-product-compliance
Lightning Source LLC
Chambersburg PA
CBHW031832170526
45157CB00001B/274